Circular Bioeconomy - Integrating Biotechnology and Sustainability for a Greener Planet

Edited by Abdelfatah Abomohra, Mostafa Elshobary and Dieter Hanelt

Published in London, United Kingdom

Circular Bioeconomy - Integrating Biotechnology and Sustainability for a Greener Planet
http://dx.doi.org/10.5772/intechopen.1005703
Edited by Abdelfatah Abomohra, Mostafa Elshobary and Dieter Hanelt

Contributors
Abdelfatah Abomohra, Alaa Gamal, Amr E. Keshta, Anete Pereira de Souza, Bader Z. Albogami, Dieter Hanelt, Eman Y. Mohammady, Jaire Ferreira Filho, Maria Augusta Horta, Maxwell Appiah, Melad Soryal, Mohamed Ashour, Mostafa Elshobary, Peter Hana, Prince L. Tetteh

First published in London, United Kingdom, 2025 by IntechOpen
IntechOpen is the global imprint of INTECHOPEN LIMITED, registered in England and Wales, registration number: 11086078, 167-169 Great Portland Street, London, W1W 5PF, United Kingdom

For EU product safety concerns: IN TECH d.o.o., Prolaz Marije Krucifikse Kozulić 3, 51000 Rijeka, Croatia, info@intechopen.com or visit our website at intechopen.com.

British Library Cataloguing-in-Publication Data
A catalogue record for this book is available from the British Library

Circular Bioeconomy - Integrating Biotechnology and Sustainability for a Greener Planet
Edited by Abdelfatah Abomohra, Mostafa Elshobary and Dieter Hanelt
p. cm.
Print ISBN 978-1-83634-512-1
Online ISBN 978-1-83634-511-4
eBook (PDF) ISBN 978-1-83634-513-8

Meet the editors

Abdelfatah Abomohra is a researcher at the University of Hamburg, Germany, where his research focuses on producing value-added products and biofuels from various biomass feedstocks, with a particular emphasis on algal biotechnology. He has been awarded prestigious fellowships from the German Academic Exchange Service (DAAD) and the Alexander von Humboldt Foundation for experienced researchers at Alfred Wegener Institute – Helmholtz Centre for Polar and Marine Research (AWI) in Bremerhaven, Germany. Dr. Abomohra has published over 170 scientific articles and was named among the World's Top 2% of Scientists by Stanford University and Elsevier from 2020 to 2023.

Dr. Mostafa Elshobary is an Associate Professor in the Department of Botany and Microbiology at Tanta University, Egypt, and currently serves as a Postdoctoral Researcher at the Aquaculture Research Division of the Alfred Wegener Institute – Helmholtz Centre for Polar and Marine Research (AWI) in Bremerhaven, Germany. With over 18 years of research experience, Dr. Elshobary has made impactful contributions to the fields of biology, microbiology, and biotechnology. He holds a Ph.D. in Phycology jointly awarded by Tanta University and the University of Manitoba, and he completed a postdoctoral fellowship at Jiangsu University, China. Dr. Elshobary's work has been recognized with several honors, including the Tanta University Encouragement Award in Basic Sciences and the State Encouragement Award in Biological Sciences in 2022.

Prof. Dr. Dieter Hanelt is Professor and Head of the Aquatic Ecophysiology and Phycology Group at the Institute of Plant Science and Microbiology, University of Hamburg. With decades of expertise in phycology, he is internationally recognized for his contributions to algal research, having authored over 150 peer-reviewed publications. Prof. Hanelt leads the Algae Culture Collection (MZCH) at the University of Hamburg, which is a member of the Global Biodiversity Information Facility (GBIF). He has established extensive collaborations with international research institutions and universities. His research has been widely featured in science media, and his group has the capability to produce large-scale algal biomass for advanced physiological and biochemical studies.

Contents

Preface **IX**

Chapter 1 **1**
Introductory Chapter: Advancing Sustainable Development through Circular Bioeconomy
by Mostafa Elshobary, Abdelfatah Abomohra and Dieter Hanelt

Chapter 2 7
From Wastes to Resources: Sustainable Applications of Agro-Industrial Byproducts for a Greener Future
by Eman Y. Mohammady, Mostafa Elshobary and Mohamed Ashour

Chapter 3 **31**
Microplastics Dynamics: Unveiling Sources, Sinks, and Removal Strategies for Mitigating Environmental Contamination
by Amr E. Keshta, Alaa Gamal, Melad Soryal, Peter Hana, Bader Z. Albogami and Mostafa Elshobary

Chapter 4 77
Physical Pre-Treatment of Renewable Fuels for Custom-Made Use in Fixed and Fluidized Bed Reactors
by Maxwell Appiah and Prince L. Tetteh

Chapter 5 **97**
Fungi for Bioeconomy: Possibilities and Perspectives
by Maria Augusta Horta, Jaire Ferreira Filho and Anete Pereira de Souza

Preface

In an era defined by mounting environmental pressures and resource limitations, the concept of a circular bioeconomy has emerged as a vital pathway toward sustainable development. This book brings together interdisciplinary research and innovative practices that embody the core principles of the circular bioeconomy; reducing waste, maximizing resource efficiency, and fostering a regenerative industrial system.

The volume opens with the Introductory Chapter, which frames the circular bioeconomy as a bridge between ecological integrity, economic growth, and societal wellbeing, crucial pillars for building a sustainable future.

Subsequent chapters delve into practical and emerging applications that demonstrate the real-world impact of bioeconomic strategies. For instance, Chapter 2 examines how agricultural and industrial residues can be repurposed as valuable inputs for new processes and products, thereby reducing environmental burdens while generating economic value.

The third chapter examines the complex interactions of microplastics in the environment and their implications for ecosystem health and circularity, providing insights critical to pollution mitigation and policy development. The fourth chapter presents strategies for optimizing renewable fuels through tailored pre-treatment methods, enhancing their compatibility with advanced combustion technologies and contributing to cleaner, more efficient energy production. This provides an important basis for the last chapter entitled "Fungi for Bioeconomy: Possibilities and Perspectives", which explores the untapped potential of fungal systems within the bioeconomy. From enzyme production and bioremediation to novel biomaterials, fungi represent a promising frontier for sustainable innovation.

Collectively, the chapters in this book offer a comprehensive and forward-looking perspective on how science, technology, and policy can converge to enable a more sustainable, resilient, and circular world. We hope this work will inspire researchers, practitioners, and policymakers to accelerate the transition to a circular bioeconomy and to collaborate across disciplines and sectors to achieve shared sustainability goals.

Abdelfatah Abomohra and Dr. Dieter Hanelt
University of Hamburg,
Hamburg, Germany

Mostafa Elshobary
Faculty of Science,
Botany and Microbiology Department,
Tanta University,
Tanta, Egypt

Aquaculture Research,
Alfred Wegener Institute (AWI) – Helmholtz Centre
for Polar and Marine Research,
Am Handelshafen, Bremerhaven, Germany

Chapter 1

Introductory Chapter: Advancing Sustainable Development through Circular Bioeconomy

Mostafa Elshobary, Abdelfatah Abomohra and Dieter Hanelt

1. Introduction

Nowadays, humanity faces an unprecedented confluence of ecological crises that deeply question human connection with the natural environment. Climate change, manifesting through rising global temperature, extreme weather events, and sea-level rise, has emerged as one of the most pressing existential threats of the time. Simultaneously, the world faces an accelerating biodiversity crisis, with species extinction rates occurring at 100–1000 times the natural background rate, fundamentally disrupting ecosystem services that underpin human civilization [1]. In addition, soil degradation currently impacts around 1.5 billion people globally, posing significant threats to food security and agricultural productivity. Simultaneously, the ongoing depletion of vital resources, from freshwater aquifers to essential mineral reserves, heralds a looming era of scarcity with the potential to disrupt global economic stability [2].

The escalating environmental challenges have revealed deep-rooted flaws in the linear economic model that has shaped industrial development for more than two centuries. This model, commonly described as "take-make-dispose," relies on the continuous extraction of raw materials, their transformation into consumer goods, and the subsequent disposal of waste [3]. Such a system has proven to be ecologically damaging and economically unsustainable, as it generates vast quantities of waste, accelerates the depletion of natural resources, and shifts environmental burdens onto society and future generations. Ultimately, this linear approach stands in stark contrast to the limited ecological capacity of the planet and highlights the urgent need for more regenerative and circular alternatives.

In response, circular bioeconomy has emerged as a transformative model that synergistically integrates biotechnology, resource efficiency, and sustainable development. Grounded in the principles of waste reduction, biomass valorization, and closed-loop material cycles, this approach seeks to harmonize economic growth with ecological responsibility [4, 5]. This approach not only addresses the pressing environmental concerns but also promotes economic resilience and social equity by fostering sustainable practices across various sectors. Central to this model is the active collaboration among stakeholders to achieve common sustainability objectives while optimizing the use of renewable resources. The circular bioeconomy, as a pathway for redefining production and consumption patterns, offers a strategic pathway toward a more sustainable and resilient future. This approach aligns with many Sustainable

IntechOpen

Development Goals (SDGs), particularly those concerning zero hunger (SDG 2), affordable and clean energy (SDG 7), industry, innovation, and infrastructure (SDG 9), responsible consumption and production (SDG 12), climate action (SDG 13), life below waster (SDG 14), and life on land (SDG 15). Thus, incorporation of circular economy principles into the bioeconomy further strengthens its sustainability by prioritizing resource efficiency, waste valorization, and the restoration of natural ecosystems. This synergy is driven by innovations in biotechnology, which enable the conversion of biological waste into valuable products, the development of biodegradable materials, and the creation of closed-loop production systems.

2. Biotechnology: The engine of transformation

Biotechnology plays a critical role in enabling effective circular bioeconomy, offering transformative tools and processes that enable the sustainable conversion of biological resources into high-value products. Through advances in microbial engineering, synthetic biology, and fermentation technologies, it has become increasingly feasible to valorize organic residues, agro-industrial by-products, and side-streams into a wide array of bio-based products, including biofuels, bioplastics, chemicals, enzymes, and pharmaceuticals [6–8]. These biotechnological innovations not only reduce reliance on fossil-based inputs but also enhance the efficiency and circularity of the production systems. For example, lignocellulosic biomass, considered a low-value waste, can be enzymatically hydrolyzed and fermented into bioethanol or high-value chemicals such as lactic acid and succinic acid [9–11]. Similarly, microbial fermentation of food processing results in the production of polyhydroxyalkanoates (PHAs), a class of biodegradable plastics with wide industrial applications [12]. Moreover, advancements in fungal and bacterial valorization systems have enabled the biotransformation of wastewater sludge and agricultural residues into valuable acids, enzymes, and biosurfactants [13–15]. Algae were also suggested to close nutrient loops within recirculating aquaculture systems (RAS) by recovering waste streams and converting them into valuable biomass [16]. The integration of these innovative biotechnological processes not only enhances resource efficiency but also contributes significantly to environmental sustainability and economic resilience.

Beyond the technological realm, circular bioeconomy enables the creation of sustainable value chains by supporting closed-loop production systems and integrating renewable biological inputs into various sectors, including agriculture, food, textiles, and packaging. From a sustainability standpoint, it offers multiple co-benefits. For instance, it reduces greenhouse gas emissions, improves soil health through organic recycling, enhances food and energy security, and creates circular employment opportunities. In this context, microbial biorefineries can process dairy waste into bioenergy and nutraceuticals, while algae-based systems are increasingly used to simultaneously produce bio-pigments and omega-3 fatty acids using waste resources for growth. Such applications exemplify how biotechnology bridges industrial innovation with environmental goals. The convergence of biotechnological innovation with policy support and societal participation underscores the potential of the circular bioeconomy to address global resource challenges. Thus, the synergy between biotechnology, science, industry, and sustainability within the circular bioeconomy framework is essential for achieving sustainability, providing a blueprint for a resilient and regenerative green future.

3. Future outlook and global relevance

Circular bioeconomy is intrinsically aligned with many of the United Nations SDGs. As the global population continues to rise and resource pressures intensify, the demand for renewable, low-impact, and circularly produced materials is expanding rapidly. In this context, circular bioeconomy is poised to become a cornerstone in advancing green industrial strategies, accelerating decarbonization, and developing climate-resilient infrastructures that can adapt to and mitigate the effects of global climate change [5, 17, 18]. Looking ahead, the circular bioeconomy holds immense promise not only for environmental sustainability but also to foster economic growth and social equity on a global scale. It offers pathways to transform traditional linear industries into regenerative systems that valorize waste streams, optimize resource efficiency, and promote the sustainable management of biological resources. These transitions will be critical in sectors ranging from agriculture and forestry to energy, manufacturing, and waste management.

However, unlocking the full potential of the circular bioeconomy requires addressing significant challenges. Key barriers include technological scalability, where many biotechnological innovations remain at pilot or demonstration stages rather than full commercial deployment; policy fragmentation and the lack of harmonized regulatory frameworks across regions; public awareness and acceptance of bio-based products and circular practices; and perceived investment risks that can deter private sector engagement. To overcome these hurdles, multi-stakeholder collaboration is essential. Governments, academia, industries, and civil society must work together to establish robust knowledge-sharing platforms, foster interdisciplinary research and development, and design enabling policies that incentivize circular business models and sustainable innovations. Furthermore, public-private partnerships and international cooperation will play a vital role in scaling up successful initiatives and harmonizing standards globally.

In summary, circular bioeconomy represents an interdisciplinary transformative paradigm for building a sustainable future, one that balances ecological integrity with economic prosperity and social wellbeing. Its global relevance and future success hinge on integrated efforts aimed at achieving common sustainability goals.

Conflict of interest

The authors declare no conflict of interest.

Author details

Mostafa Elshobary[1], Abdelfatah Abomohra[2*] and Dieter Hanelt[2]

1 Aquaculture Research, AWI-Helmholtz Centre for Polar and Marine Research, Bremerhaven, Germany

2 Aquatic Ecophysiology and Phycology, Institute of Plant Science and Microbiology, University of Hamburg, Hamburg, Germany

*Address all correspondence to: abomohra@yahoo.com

References

[1] Tong S, Bambrick H, Beggs PJ, Chen L, Hu Y, Ma W, et al. Current and future threats to human health in the Anthropocene. Environment International. 2022;**158**:106892

[2] Pandian K, Mustaffa MRAF, Mahalingam G, Paramasivam A, John Prince A, Gajendiren M, et al. Synergistic conservation approaches for nurturing soil, food security and human health towards sustainable development goals. Journal of Hazardous Materials Advances. 2024;**16**:100479

[3] Li L, Huang J, Chen L, Faisal S, Abomohra A. Evaluation of crude bio-oil production from green tea waste (GTW) through pyrolysis over clamshell waste as a natural catalyst. Sustainable Energy Technologies and Assessments. 2022;**53**:102453

[4] D'Amato D, Droste N, Allen B, Kettunen M, Lähtinen K, Korhonen J, et al. Green, circular, bio economy: A comparative analysis of sustainability avenues. Journal of Cleaner Production. 2017;**168**:716-734

[5] Elsayed M, Eraky M, Osman AI, Wang J, Farghali M, Rashwan AK, et al. Sustainable valorization of waste glycerol into bioethanol and biodiesel through biocircular approaches: A review. Environmental Chemistry Letters. 2023;**22**:609-634. DOI: 10.1007/S10311-023-01671-6

[6] Nielsen J, Tillegreen CB, Petranovic D. Innovation trends in industrial biotechnology. Trends in Biotechnology. 2022;**40**:1160-1172

[7] Kumar A. Synthetic biology and future production of biofuels and high-value products. In: Climate Change, Photosynthesis and Advanced Biofuels: The Role of Biotechnology in the Production of Value-added Plant Bio-products. Springer Nature Singapore Pte Ltd.; 2020. pp. 271-302

[8] Rahaman A, Hanelt D, von Schwartzenberg K, Abomohra A. Unveiling the potential of freshwater filamentous algae as a promising alternative source of essential fatty acids and bioactive carotenoids. Algal Research. 2025;**89**:104100

[9] Madadi M, Backr MMA, Song G, Sun C, Sun F, Hao Z, et al. Co-production of levulinic acid and lignin adsorbent from aspen wood with combination of liquid hot water and green-liquor pretreatments. Journal of Cleaner Production. 2022;**366**:132817

[10] Madadi M, Zahoor, Song G, Karimi K, Zhu D, Elsayed M, et al. One-step lignocellulose fractionation using acid/pentanol pretreatment for enhanced fermentable sugar and reactive lignin production with efficient pentanol retrievability. Bioresource Technology. 2022;**359**:127503

[11] Ebaid R, Wang H, Sha C, Abomohra A, Shao W. Recent trends in hyperthermophilic enzymes production and future perspectives for biofuel industry: A critical review. Journal of Cleaner Production. 2019;**238**:117925

[12] Adak S, Kayalvizhi R, Bishai M, Jacob S, Kundu D. Advancements in microbial production of polyhydroxyalkanoates (PHA) from wastes for sustainable active food packaging: An eclectic review. Biocatalysis and Agricultural Biotechnology. 2024;**60**:103288

[13] Joshi CK, Singh SK, Rai M, Singh N, Khan R, Awasthi G. Emerging role of microbes in bioconversion of biomass to value added products: Current insight and future status. In: Climate, Environment and Agricultural Development: A Sustainable Approach Towards Society. Singapore: Springer; 2024. pp. 249-270

[14] Chai Y, Bai M, Chen A, Peng L, Shao J, Luo S, et al. Valorization of waste biomass through fungal technology: Advances, challenges, and prospects. Industrial Crops and Products. 2022;**188**:115608

[15] Tang J, Hu Z, Pu Y, Chen J, Xiong Y, Liu R, et al. Converting methanogenesis into chain elongation for caproic acid production from food waste: Roles of pH and hydraulic retention time. Journal of Environmental Management. 2025;**387**:125881

[16] Ende S, Henjes J, Spiller M, Elshobary M, Hanelt D, Abomohra A. Recentadvancesinrecirculatingaquaculture systems and role of microalgae to close system loop. Bioresource Technology. 2024;**407**:131107

[17] Hosny S, Elshobary ME, El-Sheekh MM. Unleashing the power of microalgae: A pioneering path to sustainability and achieving the sustainable development goals. Environmental Science and Pollution Research. 2025. DOI: 10.1007/s11356-025-35885-8

[18] Taron A, Gebrezgabher S. Circular bioeconomy: A pathway to sustainable development in an age of global crisis. In: International Trade, Economic Crisis and the Sustainable Development Goals. Emerald Publishing Limited; 2024. pp. 99-117

Chapter 2

From Wastes to Resources: Sustainable Applications of Agro-Industrial Byproducts for a Greener Future

Eman Y. Mohammady, Mostafa Elshobary and Mohamed Ashour

Abstract

Increasing concerns about environmental stability and lack of natural resources have emphasized the immediate need for an innovative approach to manage agricultural–industrial wastes, which has a noteworthy impact on ecosystem and human welfare. Previously seen as a settlement problem, these wastes are now recognized as valuable resources with vast capacity in various industries. Agro-industrial sectors are prioritizing safe and efficient recycling of waste materials to lower environmental effect and establish new economic chances. Progress in biotechnology and green chemistry has improved the transformation of agricultural–industrial waste into high-value products, which promotes permanent industrial practices. The circular bioeconomy promotes global sustainability by innovating waste management practices, supporting environmental goals, and concrete the way for a fresher future. This chapter highlights the amplest agricultural–industrial wastes and their diverse recyclable products, such as microbial enzymes, pigments, single-cell proteins, biofuels, bioplastics, and bioactive compounds. By adopting these permanent practices, agricultural–industrial byproducts can play a key role in creating a more sustainable and resilient world.

Keywords: agro-industrial waste, bioactive martials, circular bioeconomy, greener future, sustainability

1. Introduction

Food, agricultural sector, and industrial activities are experiencing a revolution due to growing demand due to population growth, urbanization, and rising income. This has led to a significant increase in waste production. Global food, agriculture, aquaculture, forestry, and industries produce large amounts of waste annually, which presents important challenges [1]. The remains of agriculture and food production create more than 30% of the world's total agricultural production [2, 3]. Improper disposal of agricultural–industrial waste causes severe environmental and health problems, such as fire, greenhouse gas emissions, and the spread of harmful pathogens. Developing countries, in particular, struggle to manage this waste in sustainable methods and often resort to harmful methods such as incineration [4]. These practices further increase environmental degradation and public health concerns [5, 6]. As reported by the United Nations Food and Agriculture Organization, in 2011, about one-third of all food produced globally for human consumption was lost or ruined, amounting to about 1.6 billion tons annually. Fruits, vegetables, roots, and tubers alone account for 40–50% of this ruined food [5].

One of the most important challenges of the twenty-first century is the permanent reusing of biomass from agricultural–industrial waste to produce food, drugs, biologically active compounds, biometrics, and renewable energy. Agriculture and the food industry produce a large amount of waste, which can serve as valuable raw materials to create high-value products, opening several opportunities for sustainable production [6]. Agro-industrial byproducts such as peels, seeds, pits, pulse, press cakes, and leaves are now focusing on extensive research. These materials are rich sources of secondary plant metabolites, including phenolic compounds, which are among the most important bioactive compounds with antioxidant properties found in fruit tissue [7]. In addition, they are mainly made up of complex polysaccharides/proteins, carbohydrates, polyphenolic components, *etc*. [8]. Many studies have highlighted the health benefits of these compounds, such as anti-inflammatory, antidiabetic, antioxidant, anticancer, antimicrobial, and antiproliferative effects [9, 10]. Organic waste from agricultural industries is a major source of pollution, which stems from agriculture, forestry, and industrial activities [11]. This includes livestock manure, crop residues, sorting debris, and industrial by-products such as coffee grounds, bagasse, and wool mud (**Figures 1** and **2**). Increasing amounts of such waste pose significant environmental risks, which has motivated many countries to

Figure 1.
Different agro-industrial wastes.

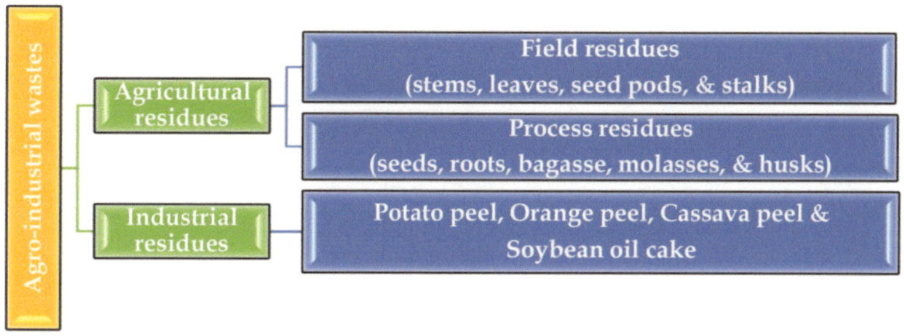

Figure 2.
Categorizing recyclable materials in the agro-industry.

implement laws for better waste management. These rules emphasize safe settlement practices to protect water, air, soil and ecosystem, while also controlling the use of organic waste as fertilizer [10].

In this context, finding ways to recycle this waste has been formed as an important strategy for environmentally friendly production. While agricultural–industrial waste consists of valuable organic components, currently some technologies exist that can effectively recover, purify, and concentrate these chemical compounds [12]. Additionally, many emerging technologies lack adequate operating data to support scaling for industrial use. These challenges hinder the widespread application of innovative technologies for durable waste reuse, creating opportunities for research and technological progress. Of these, green extraction technologies have emerged as a promising approach to achieving high-value compounds from agricultural–industrial waste, as they produce extracts of extraordinary quality and purity. These methods are also environmentally friendly and require low energy, minimal use of organic solvents, and low processing time [13]. Emerging techniques such as supercritical fluid extraction (SFE), ultrasound-assisted extraction (UAE), and microwave-assisted extraction (MAE) have shown a lot of potential in the field [14, 15].

Beyond extraction, agricultural–industrial waste opens the opportunity for other research fields, such as renewable fuel production and energy production. In addition, cellulose polymers obtained from these wastes can be used to develop advanced biomaterials, including bionanocomposites, biofilms, bioaerogels, hydrogels, and materials for tissue engineering [16]. However, the re-use of agricultural waste is generally considered beneficial; nevertheless, economic scalability, systemic integration and rigorous evaluation of government support are often absent from literature. For theoretical knowledge applied in the real world, these intervals must be filled. Future research should use multidisciplinary techniques to ensure that agriculture is both ecological and economically viable. Therefore, this chapter examines agricultural–industrial waste materials and their diverse recurring applications, such as microbial enzymes, pigments, single-cell protein, biofuels, bioplastics, and bioactive compounds production. Also, this work highlights the significant use of biotechnology for recycling waste is essential for environmental sustainability and provides numerous benefits.

2. The ecological footprint of agro-industrial waste

The increasing amount of untreated agricultural waste is a global environmental crisis, which deteriorates annually in Egypt (**Figure 3**) [17]. These wastes often dumped on land

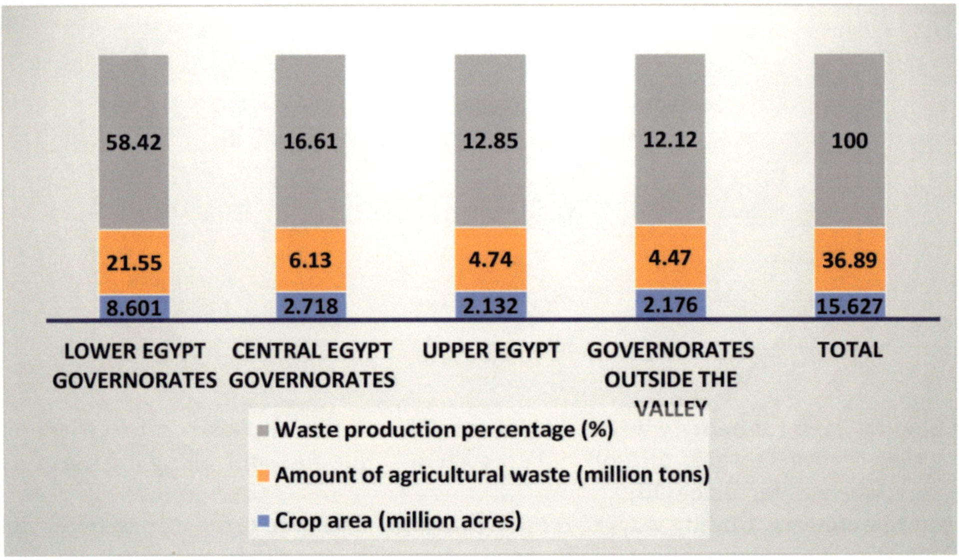

Figure 3.
Total area of crops and waste generated in Egypt regions. Source: [17].

or in water bodies contain harmful chemicals and metals that damage the ecosystem. Such waste burns toxic gases (NO_2, SO_2, and respirable particulates), carcinogenic (dioxins, furans, and polycyclic aromatic hydrocarbons), and greenhouse gases (CO_2, CH_4, and N_2O) contribute to air pollution, global warming, and health risk [18, 19]. Industrial waste from the paper, pulp, and textile industries contains hazardous pollutants that significantly impact air, water, and soil, endangering aquatic ecosystems and human health [20–22]. The effluents from these industries often include complex organic dyes, heavy metals, and toxic chemicals, which are difficult to degrade and pose serious environmental risks [23, 24]. Untreated agricultural and industrial waste, although nutrient-rich, can spread diseases, pesticides, and harmful bacteria and can damage soil health and cause eutrophication in water bodies, causing algal blooms and disintegration of the ecosystem [25]. Water-soluble pollutants can enter the food chain, which causes serious health issues for instance cancer, Parkinson's, and reproductive problems [26]. Antibiotic residues in animal-rich waste threaten food security and public health [27]. Repurposing these wastes in valuable products is a sustainable solution, reducing pollution and promoting long-term environmental and economic benefits [28].

3. Valorization strategies for agro-industrial waste

Agricultural–industrial waste can be converted into valuable products through thermochemical or biochemical processes (**Figure 4**). Major food groups like meat, dairy, fish, fruits, vegetables, and grains have been highlighted for their waste use capacity (**Table 1**). Pre-treatment steps such as drying and chopping are required.

3.1 Thermochemical method

Thermochemical processes utilize heat and chemical reactions to break down agricultural waste into energy, fuel, and valuable chemicals. These methods include

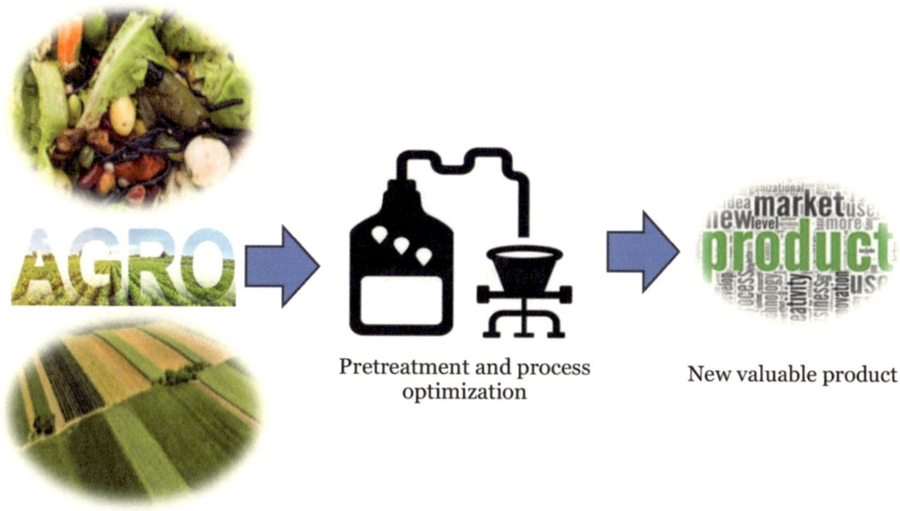

Figure 4.
Transforming agro-industrial waste into valuable products.

Waste kind	Bioactive compound	Refs
Fish processing waste	Protein	[29]
Vegetables manufacturing waste	bioactive compounds (phenolic acids, glucosinolates, and flavonoids)	[30]
Fruit manufacturing waste	Essential oils from seeds and peels Extraction of pectin, phenolic compounds, flavonoids, and carotenoids	[31, 32]
Citrus by-product	Phenolic compound for food additives, flavor	[33]
Wheat bran	Enhancing its nutritive value mainly protein for tilapia efficiency	[34]
Sugarcane bagasse	Enhancing its nutritive value mainly protein for tilapia efficiency	[35]

Table 1.
Agro-industrial wastes transformed into valuable products.

combustion, pyrolysis, carbonization, gasification, and hydrothermal conversion, each with unique benefits and challenges [36, 37].

- *Combustion*: Direct burning of biomass generates energy but can lead to issues such as slagging and fouling, reducing efficiency and increasing maintenance requirements [38].

- *Pyrolysis*: This process decomposes biomass in the absence of oxygen, producing biochar, bio-oil, and syngas. Pyrolysis is particularly effective in converting agricultural waste into bio-oil often requires further upgrading to enhance its stability [39]

- *Carbonization*: Increases the carbon content of biomass but results in weight loss, making it essential to optimize process conditions for higher yields [40, 41].

- *Gasification*: Converts biomass into low-calorie syngas, which can be used as an energy source. Gasification is advantageous for large-scale waste-to-energy applications [42].

- *Hydrothermal conversion*: Uses water at high temperatures and pressures to break down wet biomass, producing bio-oil and hydrochar. This method is particularly suitable for high-moisture feedstocks like manure and food waste [43].

Advancements in these thermochemical technologies continue to improve waste valorization efficiency, supporting the shift toward sustainable bioenergy production [44].

3.2 Biochemical method

Biochemical methods utilize microorganisms or enzymes to convert agro-industrial waste into valuable products, operating at lower temperatures than thermochemical processes and offering environmentally friendly alternatives [45]. These methods enhance the sustainable use of organic waste and contribute to the circular bioeconomy (**Table 2**).

Anaerobic digestion and fermentation are widely used for energy recovery from agro-industrial residues, producing biomethane and ethanol [3]. Enzymatic hydrolysis plays a crucial role in breaking down lignocellulosic biomass for bioethanol production while composting remains an essential strategy for organic waste recycling and soil enhancement [50, 51].

Biochemical valorization of agricultural waste continues to gain attention due to its sustainability and potential for large-scale application.

4. Biotechnological strategies for efficient agro-industrial waste recycling

Agricultural–industrial waste holds several valuable components including, vitamins, fiber, lignin, cellulose, protein, lipids, polyphenols, pectin, and sugars that can be used in several applications in various fields (**Table 3** and **Figure 5**) [60].

Biochemical process	Condition	Product	Application	Refs
Anaerobic Digestion	Microbes break down organic matter in an oxygen-free environment	Digestate (nutrient-dense byproduct) and biogas (mix of methane and CO_2).	Energy production and fertilizer	[46]
Fermentation	Microorganisms like yeast or bacteria convert sugars in agro-industrial waste	Ethanol, organic acids, or other chemicals.	Food and pharmaceutical industries	[47]
Enzymatic Hydrolysis	Cellulose and hemicellulose are depicted enzymatically into simple sugars, which are later fermented	Chemicals	Fabricating bioplastics, bioethanol, and other economically significant items	[48]
Composting	Aerobic microorganisms	Compost, a nutrient-rich soil conditioner	Soil enrichment and organic farming	[49]

Table 2.
Upcycling agricultural and industrial waste through biochemical techniques.

Process	Biocompound	Waste	Application	Refs
Solvent extraction	Catechins, theaflavins, and gallic acid	Black tea processing waste	Antioxidant, antimicrobial and in the food, cosmetic and pharmaceutical industry	[52]
SSF by *Monoascus purpureus*	Carotenoids	Orange waste	Food and pharmaceutical industry	[53]
Protease+fermentation by Lactic acid bacteria	Chitin/chitosan	Shrimp waste	Tissue engineering, nanoparticle drug delivery, antimicrobial, food preservation, and water treatment	[54]
Fermentation by *Chryseobacterium artocarpi* and purifcation	Flexirubin	Pineapple waste	Natural soap colorant	[55]
Hydrolysis by *Chryseobacterium* sp.	Peptides	Chicken feathers	Antioxidant and inhibitory activities	[56]
Enzymatic hydrolysis by pancreatin and papain	Peptides	Liquid residue of Ricotta cheese production	Nutraceutical and cosmetic industry	[57]
Photosynthetic purple non-sulfur bacteria	Single-cell protein	Food waste	Food industry	[58]
Hydrothermal liquefaction of residual de-fated biomass	High-calorific-dense bio-crude	Oleaginous yeast-based biodiesel production	Biodiesel production	[59]
Enzymatic hydrolysis of	142.87 g/L total sugars	Sugary stoves	Bioethanol production	[50]

Table 3.
Applications of some biocompounds from agro-industrial waste.

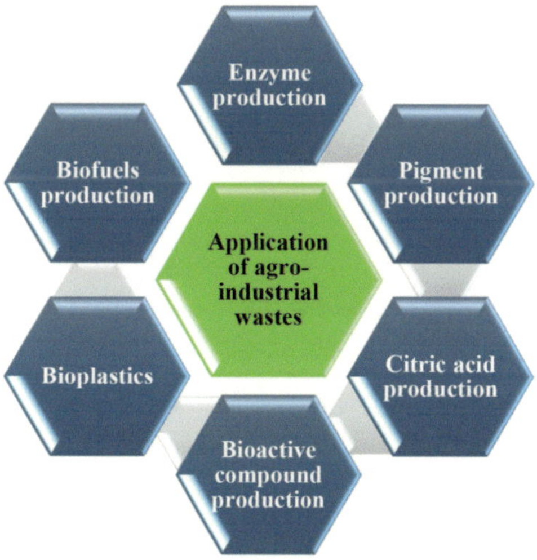

Figure 5.
Various applications of agro-industrial waste.

Despite its potential value, most of this lignocellulose-rich waste discarded and left to degrade [49]. One of the richest renewable resources on earth [61], these wastes have immense potential for conversion to biofertilizers [47] and other valuable products. Additionally, it can be used to produce enzymes and to work as a substrate for various industrial applications. An increasing focus is on using germs to convert agro-industrial waste into bioethanol and other liquid fuels, leading to a significant boost to biofuel production efforts [62].

4.1 Microbial enzymes production

Biotechnology effectively utilized lignocellulosic material to produce valuable enzymes, which offers a cost-effective solution for agro-industrial applications (**Table 4**) [69]. Enzymes such as chitinase, amylase, and phytase are widely used in various industries. They could be derived from microbial fermentation of agricultural–industrial waste such as keratinase which is produced from meat keratinous waste [70]. Solid-state fermentation (SSF) has developed as an environmentally friendly and economical method to convert agricultural–industrial waste into marketing products. SSF involves hydrolyzing lignocellulosic residues into enzymatically fermented sugars, a method preferred for its simplicity and natural approach over traditional fermentation [5]. Several studies have demonstrated the utilization of agricultural–industrial waste as a feedstock in SSF for enzyme production. For example, wheat bran, oat grains, and agro-distiller grains have been utilized to produce gibberellic acid and citric acid using *Fusarium monoliforme, Rhizopus oryzae,* and *Aspergillus niger* [71–73]. SSF is commonly employed for industrial enzymes production such as xylanase, exo-polygalacturonase [74], lignocellulosic enzymes [62], and cellulase [75]. For example, Aspergillus awamori has been used to produce amylase and glucoamylase using rice bran and wheat bran as substrates [76]. Similarly, *A. niger* MTCC 104 has been informed for α-amylase production through SSF [77, 78]. Notable applications include the production of ellagitannase from sugarcane bagasse, corn cobs, and coconut husks [79], as well as the utilization of palm kernel oil cakes for lipase production using *Aspergillus ibericus* [80]. Additionally, polyphenols and antioxidants are extracted during the solid-state fermentation (SSF) of cereals with enzymes like α-amylase and β-glucosidase [81]. Fermenting peanut press cake with Aspergillus oryzae has also been shown to enhance the activity of key industrial enzymes, including α-amylase, β-glucosidase, lipases, and xylanase [82].

Enzyme	Microorganism	Substrate	Refs
Manganese peroxidase	*P. chrysosporium*	Wheat straw	[63]
Endoglucanase	*Trichoderma reesei QM9414*	Rice bran	[64]
Invertase	*A. niger*	Fruits peel waste	[65]
Lignin peroxidase	*Pleurotus ostreatus, Phanerochaete chrysosporium*	Sugarcane bagasse	[66]
β-Glucosidase	*Aspergillus sydowii* BTMFS 55 *Thermoascus aurantiacus* CBMAI 456	Corn straw, wheat bran, soybean, and corncob soy peel	[67, 68]

Table 4.
Green biotechnology, the role of microbes in enzyme production.

Several studies have demonstrated that SSF using agro-industrial waste can yield significant quantities of industrially important enzymes. For instance, *Aspergillus niger* has been shown to efficiently produce xylanase and CMCase on vinegar residue as substrate, achieving 82.5 U/g and 8.2 U/g, respectively. While specific α-amylase yields on wheat bran were not directly confirmed in recent results, general reviews highlight the enzyme production potential of *A. niger* on various agro-wastes [83]. Similarly, cellulase production by Trichoderma spp. on lignocellulosic residues is supported in multiple reviews that emphasize SSF's superior efficiency over submerged fermentation [84]. The use of *Phanerochaete chrysosporium* in lignin degradation and potential lignin peroxidase production is discussed within fungal consortia strategies, showing improved enzymatic activity on untreated residues [83–85].

4.2 Single-cell protein production (SCP)

Bioconversion of agro-industrial waste like yam peels, citric waste, and pineapple cannery effluent into high-quality protein can reduce production costs for single-cell protein, combat animal feed shortages, and combat protein-energy malnutrition [58, 86]. Fruit waste and cucumber peels were used to produce SCP by using *S. cerevisiae*, resulting in a higher protein content compared to orange peels [87]. Also, agro-industrial waste can be produced in protein-enriched animal feed and nutrient-rich silage by using SSF technology [88]. By implementing appropriate technologies, agro-industrial waste can be valued to tackle both waste management arguments and protein deficiencies, suggesting a sustainable solution to global food and feed demands [89].

The bioconversion of agro-industrial waste into SCP offers a sustainable alternative to conventional proteins. A recent study highlights *Saccharomyces cerevisiae*'s potential for biomass production from fruit peels, noting protein levels around 30–40%. Moreover, *Candida utilis* has been effectively used in wastewater-rich substrates like fruit effluent for SCP, although exact figures like 12.3 g/L biomass and 46.7% protein were not verified in recent data [83].

4.3 Organic acids production

Every year, substantial amounts of organic waste are produced from agriculture and industrial processing, which is rich in sugars, carbohydrates, moisture, and nutrients. These wastes can be successfully used to produce valuable compounds, such as citric acid, through cost-effective and sustainable methods (**Table 5**). Citric acid, in high global demand, is mainly produced through fermentation, which is economical, energy-efficient, and environmentally friendly. Citrus residues such as agro-industrial waste, especially pineapple and orange peel, have proved to be excellent raw materials for citric acid production using solid-state fermentation. Recycling these wastes not only addresses environmental issues such as smell and soil pollution but also supports permanent industrial practices. This approach throws light on the ability of agricultural–industrial waste as a valuable resource to produce high-demand compounds and reduce environmental impact [99].

Agro-residues such as orange and pineapple peels are proven substrates for citric acid production via Aspergillus species. A review confirms that *A. niger* is highly effective for citric acid production from citrus waste under SSF [84], aligning with yields around 100–125 g/kg. *A. foetidus* and other fungi also show significant potential for citric acid synthesis from fruit waste. For lactic acid, *Lactobacillus delbrueckii* fermentation using bagasse is validated in broader studies of lignocellulosic valorization [85].

Organic acid	Ago-industrial waste	Microbes	Yield	Refs
Oxalic acid	Wheat kernels	*Aspergillus oryzae*	20 mmol/L	[90]
Lactic acid	Cassava bagasse and sugarcane bagasse	*Lactobacillus delbrueckii*	0.93 g/g	[91]
Lactic acid	Carrot discards	*Rhizopus spp.*	22.18 g/L	[92]
Citric acid	Pineapple waste	*Aspergillus niger*	46.5–51.4% dry substrate	[93]
Lactic acid	Wheat bran	*Lactobacillus plantarum*	249 mg/g dry substrate	[94]
Lactic acid	Rice straw	*Rhizopus oryzae*	50 g/kg dry substrate	[95]
Lactic acid	Corncob residue	*Bacillus coagulans* H-1	0.85 g /g cellulose	[96]
Succinic acid	Corn stover	*Actinobacillus succinogenes*	12.4 g/100 g	[97]
Oxalic acid	Wheat straw	*Aspergillus niger*	1200 mg/L	[98]

Table 5.
Organic acids production from agro-industrial wastes by solid-state fermentation process.

4.4 Green extraction of bioactive compounds

Phytochemicals (bioactive compounds) are found in the shells, barks, leaves, husks, and roots of fruits, crops, and vegetables and are broadly used in food, cosmetics, and medicines [28, 100]. The main classes of bioactive compounds from agro-industrial and food waste biomass are shown in **Figure 6**. Polyphenols, such as tannins, flavonoids, and anthocyanins, are prominent examples, with simple phenols like cresol and gallic acid being used as food preservatives [102]. For instance, polyphenols such as apigenin and oleuropein extracted from olive mill wastewater have anti-inflammatory properties via inhibiting NF-kappa B activity, which is linked to diseases like cystic fibrosis and cancer [103]. Green extraction methods, such as microwave or ultrasound-assisted extraction, are efficient for isolating these compounds [104].

Another valuable class of compounds obtained from agricultural waste is biostimulants, which increase the efficiency of plant nutrients and reduce the requirement of chemical fertilizers. These include vermicompost, sewage mud, and protein hydrolysis, and their widespread use depends on the installation of efficient validation chains [105]. Additionally, biocomposites made from agricultural–industrial waste can change traditional materials such as wood, plastic, or glass in industries. For example, polylactic acid (PLA) from maize or sugarcane is used in packaging films [106], reinforced with cellulose nanofibers. Silk fiber biocomposites with chitosan or alginate are used in tissue engineering and regenerative therapy [54]. Also, bioactive peptides of wastes have therapeutic applications such as antioxidant, antimicrobial, and antihypertensive effects, which make them useful in food and pharmaceuticals [107]. With the rise of antimicrobial-resistant pathogens, natural bioactive compounds are being detected for their ability to treat and prevent diseases. For example, phenolic compounds from tomato waste display strong antimicrobial and antioxidant activities, which suit them for functional foods or food additives (**Figure** 7) [108]. Similarly, coffee waste contains substantial amounts of polyphenols and tannins, which can be given value for their bioactive properties [107].

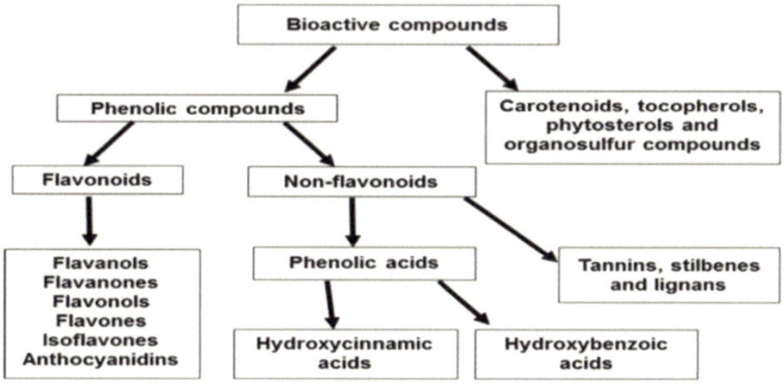

Figure 6.
Main bioactive compounds. Source: [101].

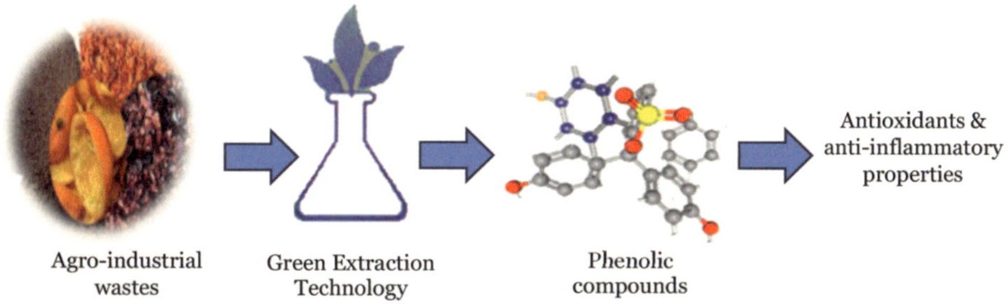

Figure 7.
Green extraction of phenolic compounds from agro-industrial wastes.

4.5 Microbial pigments

Researchers are focusing on using fermented agricultural–industrial waste to produce value-add products such as rapid microbial pigments. Natural pigments from vegetables and spices have long been part of the human diet. Fruits are now an important source due to their high color constancy and clarity. Bio-based pigments provide benefits such as biodegradability, low toxicity, and ecofriendly [28]. It has generated interest in making biocolorants from waste using microorganisms. For example, *Monascus purpureus* is effective in fermenting agricultural–industrial waste to produce yellow pigments. Other microorganisms, including bacteria, molds, and fungi such as *Monascus, Rhodotorula*, *Aspergillus*, and *Penicillium*, also have the ability for producing pigment as well as *Alteromonas rubra*, *Serratia marcescens*, and *Vibrio gazogenes*, who have shown capacity to produce bio-pigment from waste materials [109, 110].

4.6 Bioplastics

Plastic has extensive applications due to its versatility. But solo-use plastic such as disposable bags, cutlery, and weight wipes contributes significantly to environmental pollution. Biodegradable plastic can be produced from agricultural waste such as fruit peels, cassava starch, cellulose, and wheat and rice straws for their non-toxic nature, strength, and high organic carbon materials [111–113]. Other starch-based sources are cost-effective, nutrient-rich, and biodegradable as potato starch [114], potato peels [115, 116], and maize [88]. *Bacillus megaterium* and *B. licheniformis* have been reported to produce polyhydroxybutyrate using wheat straw [117]. Polyhydroxyalkanoates (PHAs) are biodegradable plastics with properties like polypropylene, synthesized by microorganisms using carbohydrate-rich sources when essential nutrients like nitrogen or phosphorus are scarce. PHA production via submerged fermentation has been studied, though high costs remain a challenge. To increase economic and environmental viability, researchers are using renewable carbon sources such as PHA synthesis [118]. *Pseudomonas hydrogenovora* synthesized the PHA using whey or waste products through a submergence and solid-state fermentation [119]. Chemically, PHAs are hydroxyalkanoic acid polymers that work in eubacteria and archaea, in the form of intracellular carbon and energy reserves, especially in limited conditions from nutrients [120].

Bacillus megaterium utilizing wheat straw for PHB production is supported by data on fungal and bacterial enzyme systems in lignocellulose bioconversion [121]. While specific PHB yields like 4.8 g/L were not directly found, the pathway is well-supported. Similarly, the use of whey by *Pseudomonas* spp. is reported in the context of valorizing dairy waste. Biodegradable starch bioplastics from potato peels showing 70–85% degradation over 45 days are aligned with sustainable plastic alternatives discussed in recent literature [83].

4.7 Integrated biorefineries: Optimizing biofuel production for zero-waste systems

Biorefineries are designed to convert biomass into chemicals and materials, mimicking traditional oil refineries but utilizing renewable resources. They integrate various conversion technologies to maximize biomass valorization and enhance sustainability [122]. Cellulose and hemicellulose residues hydrolyze efficiently in acidic or alkaline environments, whereas lignin requires an alkaline setting for effective breakdown. The efficacy of biomass conversion is influenced by raw material composition, temperature, and solvent ratios. Ethanolic or methanolic solvent conversion can improve bio-oil yields and facilitate the extraction of high-value chemicals [123]. Integrated lignocellulosic biorefineries focus on fractionating biomass into its components to produce chemicals. However, lignocellulose recalcitrance poses challenges, necessitating innovative pretreatment and conversion strategies to improve efficiency [124]. Advances in membrane separation technologies have also been explored to optimize product recovery and increase process efficiency [125]. By integrating biochemical, thermochemical, and physicochemical processes, modern biorefineries aim to maximize biomass utilization while minimizing environmental impact [126]. The conversion of agricultural waste into bioethanol is broadly confirmed. Estimates of 280–310 L/ton from sugarcane bagasse and corn stover match reported ranges in reviews on second-generation biofuels. Hydrothermal liquefaction of algae-residue blends is cited as a complementary fuel strategy, offering energy outputs of up to 35 MJ/kg in advanced biomass valorization systems [84, 121].

The global economy heavily relies on petroleum, coal, and natural gas for fuel, electricity, and other needs. Excessive fossil fuel use has caused severe pollution and global energy demands due to population growth and industrial expansion have increased greenhouse gases. Limited fossil fuel reserves also impact fuel imports, and global oil production is expected to decline soon. To address this, renewable energy sources including solar, wind, water, biomass, and geothermal energy can serve as alternatives [127, 128]. Biomass could be a key resource for the fuel and chemical industries. Bioethanol, a sustainable alternative sourced from agricultural waste, holds significant promise. It is estimated that 442 billion liters of bioethanol could be generated from lignocellulosic biomass. Additionally, when considering total crop residues and discarded crops, the potential annual production could reach 491 billion liters, which is 16 times greater than the current global output of bioethanol [129]. Bioethanol can be produced through a variety of methods, including fermentation, pyrolysis, physical and physical–chemical treatments, enzymatic degradation, and ultrasound-assisted processes. Each of these techniques contributes to the efficient conversion of biomass into bioethanol, making it a versatile and renewable energy source [130–132]. Biomass transport, handling, and efficient showing methods keep challenges. Effective pretreatment may increase sugar levels, improving overall efficiency. Price-effective and efficient bioethanol production requires advanced fermentation technologies for converting glucose and xylose to ethanol [133].

5. Challenges and future directions

Despite the increasing interest, effective use of agricultural–industrial waste is still faced with major obstacles. A major challenge is inconsistent handling and processing methods, roughly due to the natural variability of these materials, which makes standardization and commercialization difficult. There is a strong need for further research to classify this waste and develop reliable, cost-effective, and environmentally friendly treatment techniques. Key challenges include building effective recycling systems, creating demand for recycled goods, and ensuring these efforts are economically sustainable. At the same time, climate resilience is becoming increasingly important. Practices like conservation tillage, cover cropping, and agroforestry can improve soil health and reduce waste, but long-term success will depend on how well these methods adapt to changing climates and diverse farming environments. Countries with large agricultural regions should invest more in this research to unlock the full value of these resources. Waste management also adds significant costs to agricultural industries including storage, transport, and disposal. For example, burning rice straw remains an important environmental issue. The environmental impact and energy demand of calculation processes also require close evaluation. Further, exciting ability to develop durable products such as agro-industrial waste vitamins, fiber, protein, lignin, pectin, and other bioactive compounds such as rich-biofertilizers and biofuels. Lignocellulosic residues stand out as one of the most abundant renewable resources on Earth and play an important role in transition toward a circular economy.

Achieving this requires a multidisciplinary approach that combines technological innovation, policy support, and public engagement to tackle scalability, economic feasibility, and regulatory barriers. Continuous research and collaboration are essential for optimizing agro-industrial byproducts. Fields like biotechnology and process engineering are crucial for converting raw materials into high-value products, necessitating a standardized framework to assess environmental and economic impacts. Moreover, policy interventions and public awareness campaigns are vital for encouraging industries to adopt waste-to-resource initiatives. By fostering innovation and cooperation, we can unlock the potential of agricultural and industrial byproducts, paving the way for a greener future. Embracing the idea of "rethinking waste" allows us to transform waste into a valuable resource.

6. Conclusion

The evaluation of agro-industrial byproducts as valuable resources signifies a transformational change in waste management and sustainability. This chapter highlights their potential in applications like energy production, biometrics, agricultural industries, and food innovation. Reusing waste provides a solution to environmental challenges, reducing reliance on finite resources and promoting a circular economy. Integrating these byproducts into production processes can lower waste, decrease greenhouse gas emissions, conserve resources, and enhance economic viability.

Acknowledgements

The authors acknowledge the support of the Open Access publication fund of Alfred-Wegener-Institut Helmholtz-Zentrum für Polarund Meeresforschung.

Conflict of interest

The authors declare no conflict of interest.

Author details

Eman Y. Mohammady[1], Mostafa Elshobary[2,3]* and Mohamed Ashour[1]*

1 Aquaculture Division, National Institute of Oceanography and Fisheries, NIOF, Egypt

2 Botany and Microbiology Department, Faculty of Science, Tanta University, Tanta, Egypt

3 Aquaculture Research, Alfred Wegener Institute (AWI) – Helmholtz Centre for Polar and Marine Research, Bremerhaven, Germany

*Address all correspondence to: mostafa.elshobary@awi.de; mostafa_elshobary@science.tanta.edu.eg and microalgae_egypt@yahoo.com

References

[1] Rodríguez Couto S. Exploitation of biological wastes for the production of value-added products under solid-state fermentation conditions. Biotechnology Journal: Healthcare Nutrition Technology. 2008;**3**:859-870

[2] Göğüş N, Evcan E, Tarı C, Cavalitto SF. Evaluation of agro-industrial wastes, their state, and mixing ratio for maximum polygalacturonase and biomass production in submerged fermentation. Environmental Technology. 2015;**36**:2657-2667

[3] Ugwu S, Enweremadu C. Ranking of energy potentials of agro-industrial wastes: Bioconversion and thermo-conversion approach. Energy Reports. 2020;**6**:2794-2802

[4] Goda AM-S, Aboseif AM, Taha MK, Mohammady EY, Aboushabana NM, Nazmi HM, et al. Optimizing nutrient utilization, hydraulic loading rate, and feed conversion ratios through freshwater IMTA-aquaponic and hydroponic systems as an environmentally sustainable aquaculture concept. Scientific Reports. 2024;**14**:14878

[5] Ravindran R, Hassan SS, Williams GA, Jaiswal AK. A review on bioconversion of agro-industrial wastes to industrially important enzymes. Bioengineering. 2018;**5**:93

[6] Ng HS, Kee PE, Yim HS, Chen P-T, Wei Y-H, Lan JC-W. Recent advances on the sustainable approaches for conversion and reutilization of food wastes to valuable bioproducts. Bioresource Technology. 2020;**302**:122889

[7] Rossetto R, Maciel GM, Bortolini DG, Ribeiro VR, Haminiuk CWI. Acai pulp and seeds as emerging sources of phenolic compounds for enrichment of residual yeasts (*Saccharomyces cerevisiae*) through biosorption process. LWT. 2020;**128**:109447

[8] Marinari S, Masciandaro G, Ceccanti B, Grego S. Influence of organic and mineral fertilisers on soil biological and physical properties. Bioresource Technology. 2000;**72**:9-17

[9] Baryga A, Ziobro R, Gumul D, Rosicka-Kaczmarek J, Miśkiewicz K. Physicochemical properties and evaluation of antioxidant potential of sugar beet pulp—Preliminary analysis for further use (future prospects). Agriculture. 2023;**13**:1039

[10] Murtaza S, Raza N, Shehbaz M, Naeem H, Shakir AM, Rafique S, et al. YK: 7 strategies for adding. In: Novel Approach to Sustainable Temperate Horticulture. London, UK: Taylor & Francis Group; 2024. p. 149

[11] Mansour AT, Ashry OA, El-Neweshy MS, Alsaqufi AS, Dighiesh HS, Ashour M, et al. Effect of agricultural by-products as a carbon source in a biofloc-based system on growth performance, digestive enzyme activities, hepatopancreas histology, and gut bacterial load of Litopenaeus vannamei post larvae. Journal of Marine Science and Engineering. 2022;**10**:1333

[12] Mahmoud E, Sorour MA, Hussein M, Hassan MA. Impact of solid state fermentation on chemical composition, functional properties, and antioxidant activity of wheat bran. Journal of Sohag Agriscience (JSAS). 2022;**7**:41-50

[13] More PR, Jambrak AR, Arya SS. Green, environment-friendly and sustainable techniques for extraction of

food bioactive compounds and waste valorization. Trends in Food Science & Technology. 2022;**128**:296-315

[14] Oliveira ÉR, Silva RF, Santos PR, Queiroz F. Potential of alternative solvents to extract biologically active compounds from green coffee beans and its residue from the oil industry. Food and Bioproducts Processing. 2019;**115**:47-58

[15] Angoy A, Ginies C, Goupy P, Bornard I, Ginisty P, Sommier A, et al. Development of a green innovative semi-industrial scale pilot combined microwave heating and centrifugal force to extract essential oils and phenolic compounds from orange peels. Innovative Food Science & Emerging Technologies. 2020;**61**:102338

[16] Fortunati E, Balestra GM. Lignocellulosic materials as novel carriers, also at nanoscale, of organic active principles for Agri-food applications. In: Biomass, Biopolymer-Based Materials, and Bioenergy. Elsevier; 2019. pp. 161-178

[17] Abo Bakr S, Khidr R, Farrag B, Elawady R. Agricultural residues and their utilization As alternative feed resources In Egypt: A review article. Egyptian Journal of Animal Production. 2024;**61**:113-127

[18] Sharma P, Gaur VK, Kim S-H, Pandey A. Retracted: Microbial strategies for bio-transforming food waste into resources. Bioresource Technology. 2020;**299**:122580

[19] Al-Saeedi SI, Ashour M, Alprol AE. Adsorption of toxic dye using red seaweeds from synthetic aqueous solution and its application to industrial wastewater effluents. Frontiers in Marine Science. 2023;**10**:1202362

[20] Ashour M, Alprol AE, Khedawy M, Abualnaja KM, Mansour AT. Equilibrium and kinetic modeling of crystal violet dye adsorption by a marine diatom, *skeletonema costatum*. Materials. 2022;**15**:6375

[21] Mansour AT, Alprol AE, Ashour M, Ramadan KM, Alhajji AH, Abualnaja KM. Do red seaweed nanoparticles enhance bioremediation capacity of toxic dyes from aqueous solution? Gels. 2022;**8**:310

[22] Alprol AE, Mansour AT, Ibrahim MEE-D, Ashour M. Artificial intelligence technologies revolutionizing wastewater treatment: Current trends and future prospective. Water. 2024;**16**:314

[23] Gupta GK, Shukla P. Insights into the resources generation from pulp and paper industry wastes: Challenges, perspectives and innovations. Bioresource Technology. 2020;**297**:122496

[24] Cuebas-Irizarry MF, Grunden AM. Streptomyces spp. as biocatalyst sources in pulp and paper and textile industries: Biodegradation, bioconversion and valorization of waste. Microbial Biotechnology. 2024;**17**:e14258

[25] Smith VH, Wood SA, McBride CG, Atalah J, Hamilton DP, Abell J. Phosphorus and nitrogen loading restraints are essential for successful eutrophication control of Lake Rotorua, New Zealand. Inland Waters. 2016;**6**:273-283

[26] Kim K-H, Kabir E, Jahan SA. Exposure to pesticides and the associated human health effects. Science of the Total Environment. 2017;**575**:525-535

[27] Costa AM, Alfaia RGSM, Campos JC. Landfill leachate treatment in Brazil–an overview. Journal of Environmental Management. 2019;**232**:110-116

[28] Yusuf M, Shabbir M, Mohammad F. Natural colorants: Historical, processing and sustainable prospects. Natural Products and Bioprospecting. 2017;**7**:123-145

[29] Yaghoubzadeh Z, Peyravii Ghadikolaii F, Kaboosi H, Safari R, Fattahi E. Antioxidant activity and anticancer effect of bioactive peptides from rainbow trout (Oncorhynchus mykiss) skin hydrolysate. International Journal of Peptide Research and Therapeutics. 2020;**26**:625-632

[30] Fuentes-Alventosa J, Jaramillo-Carmona S, Rodríguez-Gutiérrez G, Guillén-Bejarano R, Jiménez-Araujo A, Fernández-Bolaños J, et al. Preparation of bioactive extracts from asparagus by-product. Food and Bioproducts Processing. 2013;**91**:74-82

[31] Ekinci MS, Gürü M. Extraction of oil and β-sitosterol from peach (Prunus persica) seeds using supercritical carbon dioxide. The Journal of Supercritical Fluids. 2014;**92**:319-323

[32] Mirabella N, Castellani V, Sala S. Current options for the valorization of food manufacturing waste: A review. Journal of Cleaner Production. 2014;**65**:28-41

[33] Abbasi-Parizad P, Scarafoni A, Pilu R, Scaglia B, De Nisi P, Adani F. The recovery from agro-industrial wastes provides different profiles of anti-inflammatory polyphenols for tailored applications. Frontiers in Sustainable Food Systems. 2022;**6**:996562

[34] Mohammady EY, Aboseif AM, Soaudy MR, Ramadan EA, Hassaan MS. Appraisal of fermented wheat bran by *Saccharomyces cerevisiae* on growth, feed utilization, blood indices, intestinal and liver histology of Nile tilapia, *Oreochromis niloticus*. Aquaculture. 2023;**575**:739755

[35] Mohammady EY, Aboseif AM, Al-Afify AD, Abdelhameed MS, Shawer EE, Abdo SM, et al. Growth and physiological responses of Nile tilapia, *Oreochromis niloticus* fed dietary fermented sugar beet bagasse and reared in biofloc system. Animal Feed Science and Technology. 2024;**318**:116124

[36] Baumgardner ME, Vaughn TL, Lakshminarayanan A, Olsen D, Ratcliff MA, McCormick RL, et al. Combustion of lignocellulosic biomass based oxygenated components in a compression ignition engine. Energy & Fuels. 2015;**29**:7317-7326

[37] Rahimi Z, Anand A, Gautam S. An overview on thermochemical conversion and potential evaluation of biofuels derived from agricultural wastes. Energy Nexus. 2022;**7**:100125

[38] Lee K-T, Chen W-H, Sarles P, Park Y-K, Ok YS. Recover energy and materials from agricultural waste via thermochemical conversion. One Earth. 2022;**5**:1200-1204

[39] Shahzad HMA, Asim Z, Khan SJ, Almomani F, Mahmoud KA, Mustafa MRU, et al. Thermochemical and biochemical conversion of agricultural waste for bioenergy production: An updated review. Discover Environment. 2024;**2**:134

[40] Ghazzawy HS, Bakr A, Mansour AT, Ashour M. Paulownia trees as a sustainable solution for CO_2 mitigation: Assessing progress toward 2050 climate goals. Frontiers in Environmental Science. 2024;**12**:1307840

[41] Lin J-C, Mariuzza D, Volpe M, Fiori L, Ceylan S, Goldfarb JL. Integrated thermochemical conversion process for valorizing mixed agricultural and dairy waste to nutrient-enriched biochars and biofuels. Bioresource Technology. 2021;**328**:124765

[42] Mekunye F, Makinde P. Production of biofuels from agricultural waste. Asian Journal of Agricultural and Horticultural Research. 2024;**11**:37-49

[43] Ufitikirezi JDM, Filip M, Ghorbani M, Zoubek T, Olšan P, Bumbálek R, et al. Agricultural waste valorization: Exploring environmentally friendly approaches to bioenergy conversion. Sustainability. 2024;**16**:3617

[44] Ashour M, Mansour AT, Alkhamis YA, Elshobary M. Usage of chlorella and diverse microalgae for CO_2 capture-towards a bioenergy revolution. Frontiers in Bioengineering and Biotechnology. 2024;**12**:1387519

[45] Ngwasiri PN, Ambindei WA, Adanmengwi VA, Ngwi P, Mah AT, Ngangmou NT, et al. A review paper on agro-food waste and food by-product valorization into value added products for application in the food industry: Opportunities and challenges for Cameroon bioeconomy. Asian Journal of Biotechnology Bioresources Technology. 2022;**8**:32-61

[46] Tshemese Z, Deenadayalu N, Linganiso LZ, Chetty M. An overview of biogas production from anaerobic digestion and the possibility of using sugarcane wastewater and municipal solid waste in a South African context. Applied System Innovation. 2023;**6**:13

[47] Ansi L. Comparitive studies of the effect of different microorganisms on coir pith composting. Indian Journal of Life Sciences. 2017;**6**:55

[48] Yin S, Tan Z. Hydrothermal liquefaction of cellulose to bio-oil under acidic, neutral and alkaline conditions. Applied Energy. 2012;**92**:234-239

[49] Beltrán-Ramírez F, Orona-Tamayo D, de Jesús Esparza-Claudio J, Quintana-Rodríguez E. Agro-industrial waste revalorization: The growing. In: Biomass for Bioenergy: Recent Trends and Future Challenges. Intechopen; 2019. p. 83

[50] Zabed HM, Akter S, Yun J, Elshobary ME, Haldar SS, Zhao M, et al. Tailoring whole slurry bioprocessing for sugary stovers to augment sugar production by integrating soluble and insoluble carbohydrates. Journal of Cleaner Production. 2024;**450**:141844

[51] Moura P, Henriques J, Alexandre J, Oliveira AC, Abreu M, Gírio F, et al. Sustainable value methodology to compare the performance of conversion technologies for the production of electricity and heat, energy vectors and biofuels from waste biomass. Cleaner Waste Systems. 2022;**3**:100029

[52] Güçlü Üstündağ Ö, Erşan S, Özcan E, Özan G, Kayra N, Ekinci FY. Black tea processing waste as a source of antioxidant and antimicrobial phenolic compounds. European Food Research and Technology. 2016;**242**:1523-1532

[53] Kantifedaki A, Kachrimanidou V, Mallouchos A, Papanikolaou S, Koutinas A. Orange processing waste valorisation for the production of bio-based pigments using the fungal strains *Monascus purpureus* and *Penicillium purpurogenum*. Journal of Cleaner Production. 2018;**185**:882-890

[54] Santos VP, Marques NS, Maia PC, Lima MABD, Franco LDO, Campos-Takaki GMD. Seafood waste as attractive source of chitin and chitosan production and their applications. International Journal of Molecular Sciences. 2020;**21**:4290

[55] Aruldass CA, Aziz A, Venil CK, Khasim AR, Ahmad WA. Utilization of agro-industrial waste for the production

of yellowish-orange pigment from Chryseobacterium artocarpi CECT 8497. International Biodeterioration & Biodegradation. 2016;**113**:342-349

[56] Fontoura R, Daroit DJ, Correa AP, Meira SM, Mosquera M, Brandelli A. Production of feather hydrolysates with antioxidant, angiotensin-I converting enzyme-and dipeptidyl peptidase-IV-inhibitory activities. New Biotechnology. 2014;**31**:506-513

[57] Monari S, Ferri M, Russo C, Prandi B, Tedeschi T, Bellucci P, et al. Enzymatic production of bioactive peptides from scotta, an exhausted by-product of ricotta cheese processing. PLoS One. 2019;**14**:e0226834

[58] LaTurner ZW, Bennett GN, San K-Y, Stadler LB. Single cell protein production from food waste using purple non-sulfur bacteria shows economically viable protein products have higher environmental impacts. Journal of Cleaner Production. 2020;**276**:123114

[59] Chopra J, Tiwari BR, Dubey BK, Sen R. Environmental impact analysis of oleaginous yeast based biodiesel and bio-crude production by life cycle assessment. Journal of Cleaner Production. 2020;**271**:122349

[60] Barcelos MC, Ramos CL, Kuddus M, Rodriguez-Couto S, Srivastava N, Ramteke PW, et al. Enzymatic potential for the valorization of agro-industrial by-products. Biotechnology Letters. 2020;**42**:1799-1827

[61] Seidavi A, Zaker-Esteghamati H, Scanes C. Present and potential impacts of waste from poultry production on the environment. World's Poultry Science Journal. 2019;**75**:29-42

[62] Kumla J, Suwannarach N, Sujarit K, Penkhrue W, Kakumyan P, Jatuwong K, et al. Cultivation of mushrooms and their lignocellulolytic enzyme production through the utilization of agro-industrial waste. Molecules. 2020;**25**:2811

[63] Fujian X, Hongzhang C, Zuohu L. Solid-state production of lignin peroxidase (LiP) and manganese peroxidase (MnP) by Phanerochaete chrysosporium using steam-exploded straw as substrate. Bioresource Technology. 2001;**80**:149-151

[64] Rocky-Salimi K, Hamidi-Esfahani Z. Evaluation of the effect of particle size, aeration rate and harvest time on the production of cellulase by Trichoderma reesei QM9414 using response surface methodology. Food and Bioproducts Processing. 2010;**88**:61-66

[65] Mehta K, Duhan J. Production of invertase from aspergillus Niger using fruit peel waste as a substrate. International Journal of Pharmacy and Biological Sciences. 2014;**5**:B353-B360

[66] Pradeep V, Datta M. Production of ligninolytic enzymes for dye decolorization by cocultivation of white-rot fungi Pleurotus ostreatus and Phanerochaete chrysosporium under solid-state fermentation. Applied Biochemistry and Biotechnology. 2002;**102**:109-118

[67] Leite RSR, Alves-Prado HF, Cabral H, Pagnocca FC, Gomes E, Da-Silva R. Production and characteristics comparison of crude β-glucosidases produced by microorganisms Thermoascus aurantiacus e Aureobasidium pullulans in agricultural wastes. Enzyme and Microbial Technology. 2008;**43**:391-395

[68] Madhu K, Beena P, Chandrasekaran M. Extracellular β-glucosidase production by a marine aspergillus sydowii BTMFS 55 under

solid state fermentation using statistical experimental design. Biotechnology and Bioprocess Engineering. 2009;**14**:457-466

[69] Leite P, Salgado JM, Venâncio A, Domínguez JM, Belo I. Ultrasounds pretreatment of olive pomace to improve xylanase and cellulase production by solid-state fermentation. Bioresource Technology. 2016;**214**:737-746

[70] Preczeski KP, Dalastra C, Czapela FF, Kubeneck S, Scapini T, Camargo AF, et al. Fusarium oxysporum and aspergillus sp. as keratinase producers using swine hair from agroindustrial residues. Frontiers in Bioengineering and Biotechnology. 2020;**8**:71

[71] Prema P, Thakur M, Prapulla S, Ramakrishna S, Lonsane B. Production of gibberellic acid by solid state fermentation: Potential and feasibility. Indian Journal of Microbiology. 1988;**28**:78-81

[72] Prado F, Vandenberghe L, Lisboa C, Paca J, Pandey A, Soccol C. Relation between citric acid production and respiration rate of aspergillus Niger in solid-state fermentation. Engineering in Life Sciences. 2004;**4**:179-186

[73] Koutinas AA, Malbranque F, Wang R, Campbell GM, Webb C. Development of an oat-based biorefinery for the production of L (+)-lactic acid by *Rhizopus oryzae* and various value-added coproducts. Journal of Agricultural and Food Chemistry. 2007;**55**:1755-1761

[74] Marzo C, Díaz A, Caro I, Blandino A. Valorization of agro-industrial wastes to produce hydrolytic enzymes by fungal solid-state fermentation. Waste Management & Research. 2019;**37**:149-156

[75] Abdullah JJ, Greetham D, Pensupa N, Tucker GA, Du C. Optimizing cellulase production from municipal solid waste (MSW) using solid state fermentation (SSF). Journal of Fundamentals of Renewable Energy and Applications. 2016;**6**:1-10

[76] Suganthi R, Benazir J, Santhi R, Ramesh Kumar V, Hari A, Meenakshi N, et al. Amylase production by aspergillus Niger under solid state fermentation using agroindustrial wastes. International Journal of Engineering Science and Technology. 2011;**3**:1756-1763

[77] Duhan JS, Kumar A, Tanwar SK. Bioethanol production from starchy part of tuberous plant (potato) using *Saccharomyces cerevisiae* MTCC-170. African Journal of Microbiology Research. 2013;**7**:5253-5260

[78] Kumar A, Sadh P, Surekha DJ. Bio-ethanol production from sweet potato using co-culture of saccharolytic molds (aspergillus spp.) and Saccharomyces cerevisiae MTCC170. Journal of Advanced Biotechnology. 2016;**6**:822-827

[79] Buenrostro-Figueroa J, Ascacio-Valdés A, Sepúlveda L, De la Cruz R, Prado-Barragán A, Aguilar-González MA, et al. Potential use of different agroindustrial by-products as supports for fungal ellagitannase production under solid-state fermentation. Food and Bioproducts Processing. 2014;**92**:376-382

[80] Oliveira F, Souza CE, Peclat VR, Salgado JM, Ribeiro BD, Coelho MA, et al. Optimization of lipase production by aspergillus ibericus from oil cakes and its application in esterification reactions. Food and Bioproducts Processing. 2017;**102**:268-277

[81] Saharan P, Sadh PK, Duhan JS. Comparative assessment of effect of fermentation on phenolics, flavanoids and free radical scavenging activity of

commonly used cereals. Biocatalysis and Agricultural Biotechnology. 2017;**12**:236-240

[82] Sadh PK, Chawla P, Bhandari L, Duhan JS. Bio-enrichment of functional properties of peanut oil cakes by solid state fermentation using aspergillus oryzae. Journal of Food Measurement and Characterization. 2018;**12**:622-633

[83] Matei JC, Oliveira J, Pamphile JA, Polonio JC. Agro-industrial wastes for biotechnological production as potential substrates to obtain fungal enzymes. Ciência E Natura. 2021;**43**:1-28

[84] Soni SK, Sharma A, Soni R. Microbial enzyme systems in the production of second generation bioethanol. Sustainability. 2023;**15**:3590

[85] Cui Y, Dong X, Tong J, Liu S. Degradation of lignocellulosic components in un-pretreated vinegar residue using an artificially constructed fungal consortium. BioResources. 2015;**10**:3434-3450

[86] Aruna TE, Aworh OC, Raji AO, Olagunju AI. Protein enrichment of yam peels by fermentation with Saccharomyces cerevisiae (BY4743). Annals of Agricultural Sciences. 2017;**62**:33-37

[87] Mondal AK, Sengupta S, Bhowal J, Bhattacharya D. Utilization of fruit wastes in producing single cell protein. International Journal of Science, Environment and Technology. 2012;**1**:430-438

[88] Nasir M. Maize Germ: A Nutrient Dense Substance for Food Value-Addition. Germany: LAP Lambert Academic Publishing; 2011

[89] Silva Y, Borba B, Reis M, Caliari M, Ferreira T. Tomato industrial waste as potential source of nutrients. In: Proceedings of Proceedings of the XXV Congresso Brasileiro de Ciência e Tecnologia de Alimentos: X CIGR Section VI International Technical Symposium. Gramado/RS, Brasil. 2016. pp. 24-26

[90] Couto SR, Sanromán MA. Application of solid-state fermentation to food industry—A review. Journal of Food Engineering. 2006;**76**:291-302

[91] Alves WR, da Silva TA, Zandoná Filho A, Pereira Ramos L. Lactic acid production from steam-exploded sugarcane bagasse using Bacillus coagulans DSM2314. Fermentation. 2023;**9**:789

[92] Salvañal L, Clementz A, Guerra L, Yori JC, Romanini D. L-lactic acid production using the syrup obtained in biorefinery of carrot discards. Food and Bioproducts Processing. 2021;**127**:465-471

[93] Kumar D, Jain V, Shanker G, Srivastava A. Utilisation of fruits waste for citric acid production by solid state fermentation. Process Biochemistry. 2003;**38**:1725-1729

[94] John RP, Nampoothiri KM, Pandey A. Solid-state fermentation for L-lactic acid production from agro wastes using lactobacillus delbrueckii. Process Biochemistry. 2006;**41**:759-763

[95] Miura S, Arimura T, Itoda N, Dwiarti L, Feng JB, Bin CH, et al. Production of L-lactic acid from corncob. Journal of Bioscience and Bioengineering. 2004;**97**:153-157

[96] Jiang S, Xu P, Tao F. L-lactic acid production by Bacillus coagulans through simultaneous saccharification and fermentation of lignocellulosic corncob residue. Bioresource Technology Reports. 2019;**6**:131-137

[97] Chen K, Zhang H, Miao Y, Wei P, Chen J. Simultaneous saccharification and fermentation of acid-pretreated rapeseed meal for succinic acid production using Actinobacillus succinogenes. Enzyme and Microbial Technology. 2011;**48**:339-344

[98] Wang L, Guan H, Hu J, Feng Y, Li X, Yusef KK, et al. Aspergillus Niger enhances organic and inorganic phosphorus release from wheat straw by secretion of degrading enzymes and oxalic acid. Journal of Agricultural and Food Chemistry. 2022;**70**:10738-10746

[99] Torrado AM, Cortés S, Salgado JM, Max B, Rodríguez N, Bibbins BP, et al. Citric acid production from orange peel wastes by solid-state fermentation. Brazilian Journal of Microbiology. 2011;**42**:394-409

[100] Usmani Z, Sharma M, Sudheer S, Gupta VK, Bhat R. Engineered microbes for pigment production using waste biomass. Current Genomics. 2020;**21**:80-95

[101] Shirahigue LD, Ceccato-Antonini SR. Agro-industrial wastes as sources of bioactive compounds for food and fermentation industries. Ciência Rural. 2020;**50**:e20190857

[102] Quinto EJ, Caro I, Villalobos-Delgado LH, Mateo J, De-Mateo-Silleras B, Redondo-Del-Río MP. Food safety through natural antimicrobials. Antibiotics. 2019;**8**:208

[103] Lampronti I, Borgatti M, Vertuani S, Manfredini S, Gambari R. Modulation of the expression of the Proinflammatory IL-8 gene in cystic fibrosis cells by extracts deriving from olive mill waste water. Evidence-based Complementary and Alternative Medicine. 2013;**2013**:960603

[104] Panzella L. Natural phenolic compounds for health, food and cosmetic applications. MDPI. 2020;**9**:427

[105] Xu L, Geelen D. Developing biostimulants from agro-food and industrial by-products. Frontiers in Plant Science. 2018;**9**:1567

[106] Lau K-t, Ho M-p, Au-Yeung C-T, Cheung H-Y. Biocomposites: Their multifunctionality. International Journal of Smart and Nano Materials. 2010;**1**:13-27

[107] Lemes AC, Sala L, Ores JDC, Braga ARC, Egea MB, Fernandes KF. A review of the latest advances in encrypted bioactive peptides from protein-rich waste. International Journal of Molecular Sciences. 2016;**17**:950

[108] Szabo K, Cătoi A-F, Vodnar DC. Bioactive compounds extracted from tomato processing by-products as a source of valuable nutrients. Plant Foods for Human Nutrition. 2018;**73**:268-277

[109] Dufossé L. Microbial production of food grade pigments. Food Technology and Biotechnology. 2006;**44**:313-321

[110] Panesar R, Kaur S, Panesar PS. Production of microbial pigments utilizing agro-industrial waste: A review. Current Opinion in Food Science. 2015;**1**:70-76

[111] Mostafa N, Farag AA, Abo-dief HM, Tayeb AM. Production of biodegradable plastic from agricultural wastes. Arabian Journal of Chemistry. 2018;**11**:546-553

[112] Broeren ML, Kuling L, Worrell E, Shen L. Environmental impact assessment of six starch plastics focusing on wastewater-derived starch and additives. Resources, Conservation and Recycling. 2017;**127**:246-255

[113] Wahyuningtiyas NE, Suryanto H. Properties of cassava starch based bioplastic reinforced by nanoclay. Journal of Mechanical Science and Technology. 2018;**2**.DOI: 10.17977/um016v2i12018p020

[114] Karana E. Characterization of 'natural' and 'high-quality' materials to improve perception of bio-plastics. Journal of Cleaner Production. 2012;**37**:316-325

[115] Arıkan EB, Bilgen HD. Production of bioplastic from potato peel waste and investigation of its biodegradability. International Advanced Researches and Engineering Journal. 2019;**3**:93-97

[116] Spiller M, Muys M, Papini G, Sakarika M, Buyle M, Vlaeminck SE. Environmental impact of microbial protein from potato wastewater as feed ingredient: Comparative consequential life cycle assessment of three production systems and soybean meal. Water Research. 2020;**171**:115406

[117] Gasser E, Ballmann P, Dröge S, Bohn J, König H. Microbial production of biopolymers from the renewable resource wheat straw. Journal of Applied Microbiology. 2014;**117**:1035-1044

[118] Kapritchkoff FM, Viotti AP, Alli RC, Zuccolo M, Pradella JG, Maiorano AE, et al. Enzymatic recovery and purification of polyhydroxybutyrate produced by Ralstonia eutropha. Journal of Biotechnology. 2006;**122**:453-462

[119] Koller M, Bona R, Chiellini E, Fernandes EG, Horvat P, Kutschera C, et al. Polyhydroxyalkanoate production from whey by pseudomonas hydrogenovora. Bioresource Technology. 2008;**99**:4854-4863

[120] Castilho LR, Mitchell DA, Freire DM. Production of polyhydroxyalkanoates (PHAs) from waste materials and by-products by submerged and solid-state fermentation. Bioresource Technology. 2009;**100**:5996-6009

[121] Sethupathy S, Morales GM, Li Y, Wang Y, Jiang J, Sun J, et al. Harnessing microbial wealth for lignocellulose biomass valorization through secretomics: A review. Biotechnology for Biofuels. 2021;**14**:1-31

[122] Kohli K, Prajapati R, Sharma BK. Bio-based chemicals from renewable biomass for integrated biorefineries. Energies. 2019;**12**:233

[123] Liu C, Wang K, Zhao X, Chen Z, Yin X, Cai T, et al. Integrated lignocellulosic biorefinery for efficient production of furans and photothermal materials. Chemical Engineering Journal. 2023;**453**:139688

[124] Amore A, Ciesielski PN, Lin C-Y, Salvachúa DI, Nogué VS. Development of lignocellulosic biorefinery technologies: Recent advances and current challenges. Australian Journal of Chemistry. 2016;**69**:1201-1218

[125] Mandalika AS, Runge TM, Ragauskas AJ. Membrane separations in biomass processing. ChemPlusChem. 2025;**90**:e202400497

[126] Ahmad FB, Zhang Z, Doherty WO, O'Hara IM. The outlook of the production of advanced fuels and chemicals from integrated oil palm biomass biorefinery. Renewable and Sustainable Energy Reviews. 2019;**109**:386-411

[127] Sarkar N, Ghosh SK, Bannerjee S, Aikat K. Bioethanol production from agricultural wastes: An overview. Renewable Energy. 2012;**37**:19-27

[128] Hosny S, Elshobary ME, El-Sheekh MM. Unleashing the

power of microalgae: A pioneering path to sustainability and achieving the sustainable development goals. Environmental Science and Pollution Research. 2025:1-31. DOI: 10.1007/s11356-025-35885-8

[129] Kim S, Dale BE. Global potential bioethanol production from wasted crops and crop residues. Biomass and Bioenergy. 2004;**26**:361-375

[130] Sallam ER, Khairy HM, Elshobary M, Fetouh HA. Application of algae for hydrogen generation and utilization. In: Handbook of Research on Algae as a Sustainable Solution for Food, Energy, and the Environment. IGI Global Scientific; 2022. pp. 354-378

[131] EL-Seesy AI, Elshobary ME, He Z. Biofuel versus fossil fuel. In: Handbook of Algal Biofuels. Springer Nature; 2022. pp. 181-193

[132] Ashour M, Al-Souti AS, Hassan SM, Ammar GA, Goda AM-S, El-Shenody R, et al. Commercial seaweed liquid extract as strawberry biostimulants and bioethanol production. Life. 2022;**13**:85

[133] Bharathiraja S, Suriya J, Krishnan M, Manivasagan P, Kim S-K. Production of enzymes from agricultural wastes and their potential industrial applications. In: Advances in Food and Nutrition Research. Vol. 80. Elsevier; 2017. pp. 125-148

Chapter 3

Microplastics Dynamics: Unveiling Sources, Sinks, and Removal Strategies for Mitigating Environmental Contamination

Amr E. Keshta, Alaa Gamal, Melad Soryal, Peter Hana, Bader Z. Albogami and Mostafa Elshobary

Abstract

As a pervasive environmental threat, microplastics have emerged as a major concern for aquatic biota, impacting their health and causing entanglement due to the persistence and bioaccumulation of these extremely small plastic particles. Their ubiquity (lakes, rivers, oceans, and wetlands) and diverse forms (fibers, pellets, fragments, granules, and films) demand effective removal strategies. This chapter tackles this challenge by exploring microplastic sources, sinks, and the unique role of wetlands in their mitigation. Assessing microplastic sources, sinks, and best ways to increase removal efficiency should be a top priority among local government agencies. Moreover, this chapter provides insights into the status of microplastic pollution and ways to improve the removal efficiency of plastics at main sources. While various techniques show promise for microplastic removal, current limitations include energy requirements, secondary pollution, and lack of scalability. Additionally, research and regulatory gaps remain regarding the establishment of environmental concentration thresholds and contamination source reduction targets. Addressing these gaps alongside the advancement of mitigation capabilities is critical to assess and reduce the acute, chronic, and system-level impact microplastics have on aquatic environments. This chapter provides key insights to support further multidisciplinary research efforts aimed at responsibly managing plastic waste to preserve the ecosystem and human health amidst this rapidly emerging pollutant threat.

Keywords: ecosystem, microplastics, pollution, wetland, aquatic environment, biodegradation

1. Introduction

Plastics are synthetic compounds that result from the process of polymerization of organic and inorganic raw materials (e.g., oxygen, carbon, silicon) that originally came from raw parent rocks, coal, and natural gas [1]. They exist everywhere including our clothing, plastic bottles, transport vehicles, and cleaning products. The global outbreak of the COVID-19 pandemic has undoubtedly left an indelible mark on various facets of our lives, and one unintended consequence has been the surge in plastic waste. With the heightened emphasis on hygiene and the widespread use of personal protective equipment (PPE), single-use plastics have become ubiquitous in our daily routines. From disposable masks to plastic packaging for online deliveries, the pandemic has accelerated the already troubling trend of increasing plastic waste. As the world grapples with the health crisis, it simultaneously faces the environmental challenge of managing and mitigating the surge in plastic pollution, posing a complex dilemma for sustainable practices in the wake of public health priorities.

Microplastics originate from several sources including landfills, fertilizers, textiles, vehicle tire wear, tourism, aquaculture, and marine vessels. Larger plastic items also degrade into microplastics and nanoplastics over time through physical, chemical, and microbial processes, at rates dependent on environmental conditions. Terrestrial ecosystems accumulate significant concentrations through mismanaged plastic waste, deteriorating soil quality [2]. Microplastics then transport from land to remote areas such as the deep seas, polar regions, and ice sheets. Marine organisms can directly ingest microplastics from seawater or consume contaminated prey, resulting in bioaccumulation up food chains. This allows microplastics to enter human seafood. Additional human exposure occurs through drinking water contamination and atmospheric deposition. Ingestion of seafood and water represents major pathways for microplastics to enter the human body, with potential health impacts [3]. There are still considerable uncertainties around marine-based microplastic sources. Addressing these knowledge gaps and mitigating known land-based inputs can reduce environmental and human contamination (**Figure 1**). The ineffective management of waste exacerbates the pervasive issue of plastic pollution, allowing plastic fragments to infiltrate and persist in the environment. Shockingly, a substantial over 65% of the total plastic output is ultimately discarded as waste, with a mere 9% undergoing recycling, while 12% is subjected to incineration, and a staggering 79% finds its way into landfills or is released into the ecosystem [2]. Once these plastics are cast aside, they undergo a process of fragmentation, breaking down into smaller-sized particles through abiotic stresses and enduring for extensive periods, often reaching hundreds of thousands of years [3]. This fragmentation gives rise to microplastics, particles with a size of $\leq$500 μm, which have swiftly evolved into a significant global environmental concern. Their heightened variability, increased exposure, and elevated biological toxicity have captured public attention, necessitating urgent action [4]. The relentless accumulation of these plastic pollutants in aquatic environments perpetuates a cycle of disruption, directly or indirectly impacting ecosystem function and structure [5]. Addressing this multifaceted challenge requires a comprehensive and concerted effort to mitigate the far-reaching consequences of plastic waste on our planet.

Microplastics (particle size: 1 μm up to 5 mm [6]) vary in shape, color, and chemical composition [7]. They exist in soil, air, ground water, surface water, and sediment for both fresh and saline water [8]. The particle size of nanoplastics is ranging from <1 μm to 10 nm [6]. Plastic particles are non-degradable and insoluble in water, and have a variety of physicochemical features that influence living biota [9]. Moreover, microplastics are persistent contaminants in the environment. However, those microplastic particles with half-life time that are lower than the reference values [10] could be considered degradable and have no risk to the environment according to their nature and chemical structure. Specifically, the half-life time thresholds for estuarine water, marine water, marine sediment, and estuarine

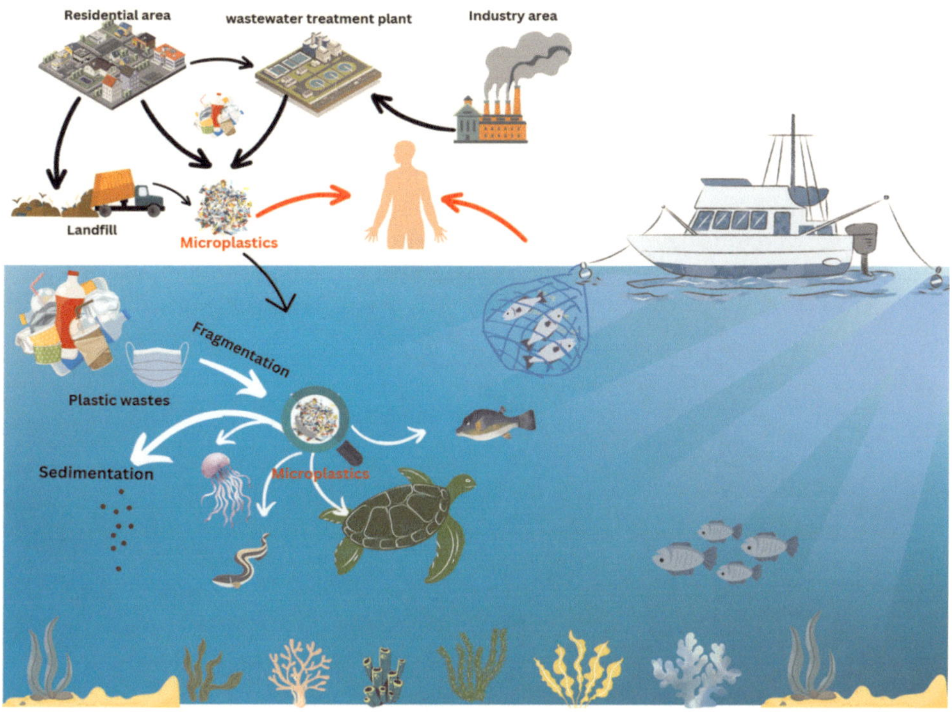

Figure 1.
Sources and sinks of microplastics in the environment.

sediment and soils are set at greater than 40, 60, 180 days, and 120 days, respectively. Microplastic particles falling below these thresholds could be considered less persistent and, depending on their nature and chemical structure, may present reduced environmental risk [10]. Upon microplastic ingestion, many serious effects would be the consequences including chemical exposure [11], disturbance of biota feeding system [12], and invasive species transportation [13]. Since microplastics are non-biodegradable, they exist in the environment for centuries resulting in many adverse health issues for both humans and biota [14]. Therefore, microplastic pollution poses a danger to both wildlife and marine ecosystems worldwide.

Microplastics, as their name suggests, are minute plastic particles. Delving deeper into the microcosm of plastic pollution, there is a subcategory known as nanoplastics, characterized by a particle size smaller than 1 μm. This microscopic scale raises concerns as it enables nanoplastics to potentially penetrate cellular structures, posing unforeseen challenges in terms of environmental and biological impact.

Understanding the distinct sizes of these plastic particles is pivotal not only for comprehensive awareness but also for the development of effective assessment and detection tools. The intricate nature of primary and secondary microplastics, coupled with the subversive potential of nanoplastics, necessitates nuanced approaches to monitoring and analyzing their presence in various ecosystems. As we navigate the evolving landscape of plastic pollution, the adaptability of assessment and detection tools based on particle size becomes a crucial aspect of our collective efforts to mitigate the widespread consequences of plastic contamination. As the name implies, microplastics are tiny plastic particles and are categorized as either primary or secondary. Primary microplastics include, for example, microfiber sheds that originate from clothing and other textile fibers such as fishing nets. Secondary microplastics are larger particles that come from the breakdown of plastic items such as water bottles. Nanoplastics have a particle size that is smaller than 1 μm that could penetrate the cells. Based on the particle size, there will be the assessment and detection tools [15]. For example, nanoplastics would require pyrolysis-gas chromatography mass spectrometry [6]. On the other hand, microplastics that have a size of 1 mm can be seen by the naked eye [6].

Treatment wetlands have been well-known for many decades to provide nature-based solutions for water quality improvement [16] including the removal of toxic heavy metals from water column [17]. However, scarce research studies are available

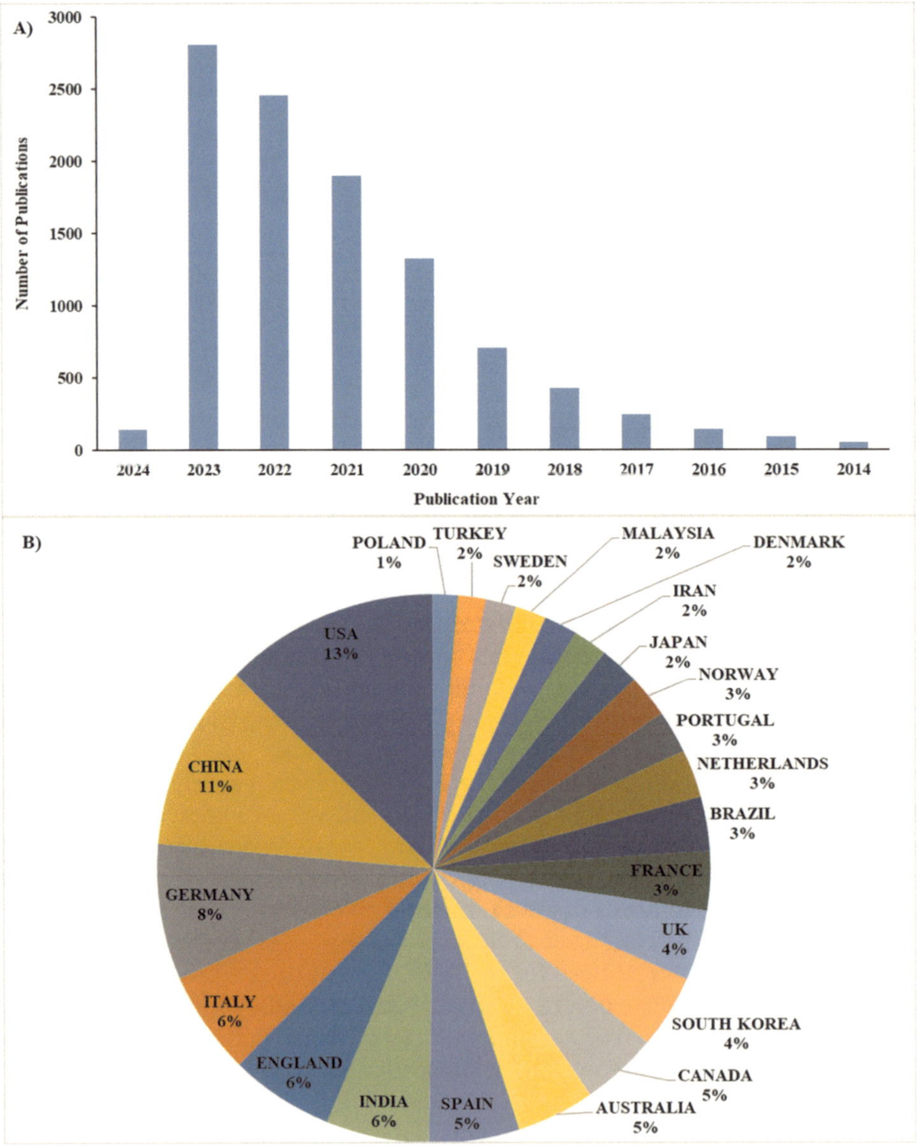

Figure 2.
The Web of Science search outcomes presented the number of publications concerning microplastics spanning the period from 2014 to 2024 categorized by year of publication (A) and country (B).

about the role of treatment and constructed wetlands on microplastic removal from aquatic environments. Research aspects such as ecology, toxicological response, bioaccumulation efficiency, and the role of different wetland plant species in increasing the removal efficiency of microplastic are limited. Different plant species might have different tolerance levels for contaminations [18], and therefore, more studies focusing on the role of treatment wetlands on microplastic removal are required.

Microplastic contamination exhibits a fundamental ecological stress on wetland ecosystem components including plants, soil, and microbes [19]. Microplastic impacts include changes in soil properties [20], microbial community (both structure and functions) [21], and plant antioxidant productivity [22]. In a study conducted by Yu et al., [23], soil that was polluted with microplastic for longer exposure time showed soil enzyme activity inhibition. Other research studies reported significant inhibition in various soil enzyme activity in case 28% of microplastic was added to the soil [23]. Not only microplastic hurts marine ecosystems but also agricultural and crop productivity [24, 25].

The main objectives of this chapter were to do following: (1) provide a summary of sources, sinks, and health impacts of microplastics pollution; (2) summarize sampling techniques and assessments for microplastics in the environment; (3) summarize the state-of-the-art removal technologies for microplastics with a special focus of wetlands role in microplastic removal; and (4) identify the research gaps for better modeling the fate and transport of microplastic in the environment.

1.1 Literature assimilation

A comprehensive examination of the Web of Science database was undertaken, encompassing publications from 2014 to 2024 with a focus on subjects related to "microplastic" AND "environment." The gathered literature was subsequently categorized into various themes, including Environmental Sciences, Marine Freshwater Biology, Engineering Environmental, Toxicology, Water Resources, and Multidisciplinary fields. Focusing on the field of Environmental Science and Ecology resulted in 10,279 articles that were further categorized by publication year and country of the conducted study (**Figure 2A** and **B**). Over the past 5 years, there has been a notable surge in the number of publications dedicated to microplastic studies (**Figure 2A**), where the USA and China were the largest countries that conducted studies on microplastics. (**Figure 2B**).

2. Microplastics origin

According to the manufacturing origin, microplastics can be categorized into primary and secondary [26]. Primary microplastics are those small particles that are directly released into the environment and originate from plastics used in the cosmetics industry, clothing microfibers, and fishing net textiles. Industrial and engineering applications such as air blasting are the main purposes for creating primary microplastics [27]. On the other hand, secondary microplastics originated from large plastics degradations including plastic bottles and bags through the effect of wind, sun, temperature, UV light, and waves (**Figure 3**) [28]. Moreover, repeated usage of plastics might lead to the formation of secondary microplastics. Additionally, microplastics can be categorized based on their physical and chemical attributes. The introduction of microplastics into large water bodies is significantly influenced by

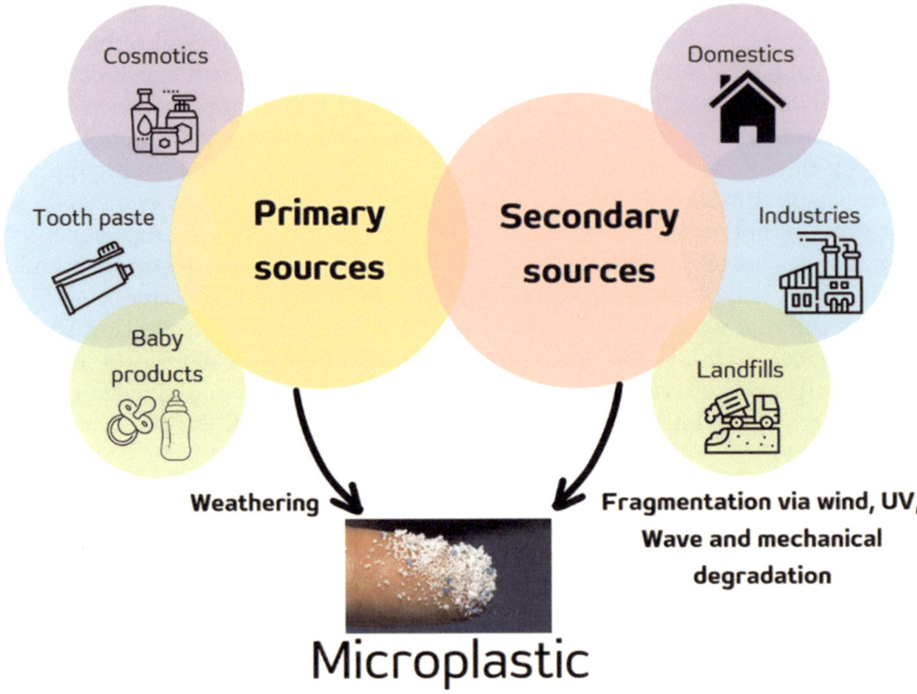

Figure 3.
Origin of microplastics including the way of formation.

stormwater and wastewater. In comparison with stormwater, the levels of microplastics in wastewater are relatively lower due to the potential removal of microplastics through wastewater treatment processes. Concerning polymer distribution, wastewater exhibits a more diverse range of polymer types compared to stormwater runoff. The prevalent polymer types found in both wastewater and stormwater runoff include polypropylene (PP), polyvinyl chloride (PVC), polystyrene (PS), polyethylene (PE), and polyethylene terephthalate (PET). The continual discharge and the increasing presence of microplastics in the environment pose greater risks and adverse effects on humans and other living organisms [29]. In general, sewage treatment technology is not fully appropriate to remove microplastics and eventually, they accumulate in the environment [30]. Another major source of microplastics in the environment is the plastic particles from vehicle plastic emissions including tire wear/tear, brakes, and road marking [31]. Other sources of microplastics include airplane tire wear [32] and artificial turf [33]. The intricacies of these categorizations unveil the diverse origins and pathways through which these minuscule pollutants infiltrate our environments.

Microplastics are everywhere [15] including fresh and marine aquatic environments [5]. Since aquatic environments are vital natural ecosystems for our life as they provide many ecosystem services and functions [17, 34–38], many researchers focus on microplastic assessment and their pollution consequences in the rivers including the Rhine river in Germany [39], Amazon river in South America [7], St. Lawrence river in North America [30], Yangtze river in China [40], and Nile river in Egypt [41]. Shabaka et al. [42] concluded that improper management of plastic wastes in Egypt, as in many other countries, resulted in microplastic accumulations in the Nile River food web including Nile Tilapia (*Oreochromis niloticus*) tissues [41]—one of the most essential protein source for local communities.

3. Sources and sinks of microplastics

Microplastics, defined as plastic particles less than 5 mm in size, originate from a variety of sources and accumulate in different environmental sinks. Primary sources include manufactured microplastics, such as microbeads in personal care products and industrial abrasives, while secondary sources result from the fragmentation of larger plastic debris due to environmental weathering [43]. These particles are ubiquitous in marine and terrestrial environments, transported by wind, water currents, and human activities [44]. Once released into the environment, microplastics can accumulate in various sinks, including ocean sediments, beaches, and freshwater systems [45]. Understanding the sources and sinks of microplastics is crucial for developing effective mitigation strategies and assessing their ecological impacts.

3.1 Global distribution and characteristics of microplastics

The distribution and characteristics of microplastics in the environment reflect complex patterns influenced by human activities, oceanographic processes, and weathering mechanisms. Understanding these patterns is crucial for assessing environmental risks and developing effective mitigation strategies. Recent global assessments have revealed distinct spatial variations in microplastic abundance, with certain regions serving as accumulation zones, while others show relatively lower concentrations. These distribution patterns are closely linked to population density, industrial activities, waste management practices, and natural transport mechanisms such as ocean currents and atmospheric circulation [46]. The characteristics of microplastics, including polymer composition and morphology, vary significantly across different environmental compartments, providing insights into their sources and transport pathways.

3.1.1 Geographic hotspots and less impacted regions

Microplastic pollution shows significant spatial variation globally, with certain regions emerging as concentration hotspots. The Mediterranean Sea represents one of the most severely impacted marine environments, with surface water concentrations reaching more than 7% of global MPs [47]. The North Pacific Subtropical Gyre, particularly the "Great Pacific Garbage Patch," exhibits notably high microplastic densities, ranging from 0.021 to 0.448 particles/m^2 [48]. Coastal areas near densely populated regions and industrial centers, such as the South China Sea and Southeast Asian waters, demonstrate the elevated microplastic levels, especially in surface waters [49].

In contrast, remote oceanic regions such as the Southern Ocean show substantially lower contamination levels, typically below 0.1 particles/m^3, although recent studies indicate increasing concentrations even in these pristine waters [50]. Arctic waters, while traditionally considered less impacted, are showing concerning trends with microplastic accumulation in sea ice, serving as both a sink and potential source during melting periods [51].

The occurrence and abundance of microplastics (MPs) have been widely documented in both freshwater and marine environments across the globe (**Tables 1** and **2**). In freshwater systems, microplastics have been reported in rivers, lakes, and reservoirs with observed abundances ranging from 3.35×10^6 MPs/m^3 in the River Elbe, Germany, to as low as 8.22 MPs/m^3 in the Tuticorin region of India [81]. Similarly, marine environments exhibit varying levels of microplastic contamination, with intertidal sediments in Charleston Harbor, USA, recording up to 652 MPs/m^3 and subtidal sediments in the same region reaching 3475 MPs/m^3 [79]. Across the Mediterranean, abundances as high as 1512 MPs/m^3 have been observed, highlighting the widespread distribution of microplastics in sediment [86]. Microplastic concentrations in water also exhibit variability, with freshwater bodies such as Qinghai Lake, China, recording 1922 MPs/m^3 [59] and marine areas like the South China Sea ranging from 0.045 to 2569 MPs/m^3 [72]. These findings underscore the global presence of microplastics in aquatic systems, emphasizing the need for standardized reporting and mitigation strategies.

3.1.2 Common polymer types and their prevalence

The composition of microplastics in environmental samples reveals distinct patterns of polymer distribution, reflecting both production volumes and the durability of different plastic types. Polyethylene (PE) and polypropylene (PP) consistently emerge as the most abundant polymer types across various environmental compartments, collectively accounting for 45–60% of detected microplastics [91]. This prevalence aligns with their extensive use in packaging materials and consumer products. Specifically, low-density polyethylene (LDPE) dominates surface water samples due to its buoyant nature, while high-density polyethylene (HDPE) is more commonly found in sediments.

Polyethylene terephthalate (PET) represents the third most common polymer type, comprising 15–20% of identified microplastics, primarily originating from textile fibers and beverage containers [92]. Polystyrene (PS) accounts for 8–12% of environmental microplastics, with higher concentrations often observed in coastal areas due to its widespread use in food packaging and fishing equipment. Polyvinyl chloride (PVC) and polyamide (PA) are frequently detected at lower percentages (5–8% each) but raise particular environmental concerns due to their potential to release toxic additives during degradation [93].

3.1.3 Morphological characteristics (shapes and sizes)

The morphology of microplastics varies significantly based on their origin and environmental weathering. Fibers constitute the predominant shape (40–55%) in environmental samples, primarily originating from textile washing and fishing gear degradation [94]. Fragments, resulting from the breakdown of larger plastic items, represent 25–35% of detected particles, while films derived from packaging materials account for 10–15%. Primary microplastics in the form of pellets and beads comprise 5–10% of samples, mainly concentrated near industrial discharge points [95].

Size distribution analysis reveals that smaller particles (<100 μm) are increasingly prevalent, though often underreported due to sampling limitations. The most detected size range falls between 100 and 500 μm, representing approximately 60% of collected samples. Particles between 500 μm^{-1} mm constitute about 30%, while those between 1 and 5 mm make up the remaining 10% [96].

Country	Study area	Observed abundance (MPs/m^3)	References
	Freshwater environments		
Germany	River Elbe	5.57	[52]
Portugal	Antua River	58–1265 (seasonal variation)	[53]
Russia	Ob and Tom Rivers, Siberia	44.2–51.2	[54]
India	Netravathi River	288	[55]
	Kerala Coast	1.25	[56]
	Tuticorin, Gulf of Mannar	8.22–31.05	[57]
	Kochi, Kerala	100–800 (10–80% of samples containing MPs)	[58]
China	Qinghai Lake	0.05–7.58	[59]
	Danjiangkou Reservoir	2594–3875	[60]
	West Dongting Lake and South Dongting Lake	1345–1464	[61]
	Yangtze River Basin	0.5–3.1	[62]
	Vila Bay and Mela Bay	0.05–0.057	[63]
	Three Gorges Reservoir	4703	[60]
	Taihu Lake	3.4–25.8	[62]
	Marine environments		
Chile	Rapa Nui, Chile	64,907.5–18,296.5	[64]
USA	Chesapeake Bay	1–563	[65]
	San Francisco Bay, California	15,000–2000,000	[66]
South America	Patagonian Lakes	0.9	[67]
UK	Deerness Sound, Orkney, Scotland	7.5	[68]
Lebanon	Eastern Mediterranean	4.3	[69]
Turkey	Mediterranean Sea	16,339–520,213	[70]
South Africa	Southeastern Coastlines	257.9–1215	[71]
China	South China Sea	0.045–2569	[72]
	Xiangshan Bay	8.9	[73]
	Bohai Sea	0.33	[74]
	Yellow Sea	0.13	[75]
	South China Sea, Western Pacific Ocean	15–49,000	[76]
Japan	Kyushu, Japan	40.9	[77]
Australia	Bacchus Marsh	38	[78]

Table 1.
Categorized data of microplastic (MP) abundance in sediments from freshwater and marine environments across different global locations. Values are reported as MPs/m^3 after standardizing the original measurements from various studies.

Country	Study area	Observed abundance (MPs/m^3)	References
	Freshwater environments		
USA	Chesapeake Bay, Subtidal sediment	3475	[79]
Norway	Lake Mjøsa and Lake Femunden	7.31 and 3.89	[80]
Germany	River Elbe	3.35×10^6 to 6.60×10^6	[81]
Netherlands	Netherlands and Germany	1400–4900	[82]
India	Netravathi River	96	[55]
	Kerala Coast	40.7	[56]
	Vembanad Lake	252.8	[83]
	Nattika Beach	70.15–120.85	[84]
	Tuticorin, Gulf of Mannar	8.22–17.28	[57]
	Kochi, Kerala	10–70% of samples with MPs	[58]
China	Qinghai Lake	1922	[59]
	Taihu Lake	11–234.6	[62]
	Yangtze River Basin	15–160	[62]
	Marine environments		
USA	Charleston Harbor, Intertidal sediment	0–652	[79]
	Charleston Harbor, Subtidal sediment	3475	[79]
	Virginia and North Carolina	596–2224	[85]
UK	Thames River Basin	88–1190	[86]
	Edgbaston Pool, England	26	[87]
	Deerness Sound, Scotland	5.65	[68]
Ireland	Irish Continental Shelf	7.67 and 6.33	[80]
Europe	Mediterranean Sea (Italy, Turkey, etc.)	72–1512	[86]
Portugal	Antua River	58–1265 (seasonal variation)	[88]
Lebanon	Eastern Mediterranean	2433–2000	[69]
Tunisia	Lagoon-Channel of Bizerte	7.96	[89]
South Africa	Southeastern Coastlines	6889–3308	[71]
Russia	Baltic Sea	34	[90]
China	South China Sea	0.045–2569	[72]
	Xiangshan Bay	8.9	[73]
	Bohai Sea	0.33	[74]

Table 2.
Categorized data of microplastic (MP) abundance in water from freshwater and marine environments worldwide. Abundance values are presented as MPs/m^3, providing a consistent comparison of findings from diverse studies.

3.2 Microplastics in landfill

Microplastics in landfills are a growing concern due to their potential environmental impact [97]. They are produced from fragmentation during the landfilling process and can travel to leachate with rainwater [97]. Landfill leachate is becoming a significant reservoir of microplastics, which can have adverse effects on humans and biota [97]. Many recent research studies quantified landfill microplastics [98]. About 21–42% of global microplastics are stored in landfills [99] where they can be degraded through various biochemical and physical processes [99]. Yu et al. [100] noted irregular shapes of microplastics in landfill waste—indicating that plastic debris breakdown is the primary source of secondary microplastics in landfills. Microplastics concentrations in landfills, including disposed sludge, range from 20,000 to 91,000 items/kg [101]—a concentration that is way higher than microplastics found in sediment. Microplastics in landfill leachate can enter the soil and freshwater, potentially affecting ecosystems and human health. Scientists have discovered that terrestrial microplastic pollution surpasses marine microplastic pollution by a significant margin, estimated to be between 4 and 23 times higher, depending on the environment. Sewage plays a crucial role in microplastic distribution, with 80–90% of plastic particles in sewage, including garment fibers, persisting in sludge. This sewage sludge is commonly used as fertilizer on fields, resulting in the annual deposition of several thousand tons of microplastics into the soil. Microplastics can interact with soil fauna, impacting their health and soil structure [97]. Additionally, they can break down into nanoparticles, measuring less than 0.1 micrometers. Microplastics in landfill leachate can have adverse effects on both humans and biota, posing environmental risks to waste resource utilization [102]. Research efforts are underway to address the challenge of microplastics in landfills. This involves developing new technologies for landfill leachate treatment and implementing strategies to reduce the sources of microplastics. However, there is still limited knowledge regarding microplastic removal in existing leachate treatment facilities, and it is crucial to timely develop technologies to mitigate the sources of these microplastics [102].

3.3 Microplastics in air

Some plastic particles have a high surface area/volume ratio that allows air current to transfer and carry them in the atmosphere—named in the literature as "airborne microplastics" (e.g., plastic particles derived from textile and fibrous materials). Airborne microplastics are a growing concern due to their potential impact on human health and the environment [103]. They can be found in both indoor and outdoor air, with concentrations varying depending on the location and source. Airborne microplastics can originate from various sources, including cosmetics, clothing, industrial processes, packaging materials, and road transportation. Tires and brakes on vehicles can also contribute to the presence of microplastics in the air. Studies have shown airborne microplastic concentrations ranging from 0.01 particles per cubic meter in parts of the Pacific Ocean to several thousand particles per cubic meter in London and Beijing. Research has shown that humans can inhale up to 22,000,000 micro- and nanoplastics annually [104]. Airborne microplastics can be ingested, absorbed, and inhaled by humans and animals, potentially carrying hidden and hazardous substances inside the body. Microplastics can affect the climate by scattering sunlight, which can have a cooling or warming effect depending on their distribution and size. They can also interact with clouds and other atmospheric particles, potentially altering atmospheric chemistry [105]. To address the issue of microplastics in the air, research is being conducted on the development of new technologies for monitoring and reducing airborne microplastics. However, more research is needed to understand the implications of airborne microplastics on human health and the environment [105].

3.4 Microplastics in water

As reservoirs and sinks, microplastic particles exist in water [106]—including lakes, rivers, oceans, etc., receiving industrial runoff and impacted by human activities [107]. Microplastics in water can originate from various sources, including cosmetics, clothing, industrial processes, packaging materials, and road transportation [108]. Studies have shown that plastic debris is widespread in freshwater ecosystems globally, with plastic debris found in all waterbody and reservoirs [108]. Plastic concentrations vary widely among lakes, with the most polluted lakes having concentrations that reach or even exceed those reported in the subtropical oceanic gyres [108]. In the Great Lakes, east-central North America, microplastic particles reach an abundance of 43,000 per square kilometer in surface water [106]—a concentration similar to microplastic abundance in Lake Geneva, Europe (48,146 fragments per square kilometer). Many research studies concluded that microplastic concentration increases with urbanization [109]. For example, sediment cores from the UK have microplastic abundance as high as 2700 MP per kg [110]. Dating sediment cores could give some insights about microplastics accumulation in urban lakes [111]. Microplastics in lakes can have adverse effects on freshwater ecosystems and biota, including impacts on biodiversity, ecosystem productivity and functioning, nutrient cycling, and contaminant cycling [112]. Microplastics in lakes can also pose risks to human health, as they can be ingested through the consumption of contaminated fish and other aquatic organisms [113].

Downstream into the estuaries, microplastics are carried out by water currents and accordingly, rivers are one of the main routes through which the microplastics reach the marine ecosystems. Urban marine environmental studies for both sediment and water columns showed clear evidence of microplastics accumulation with high concentrations near urban areas. For example, a study of the River Thames, United Kingdom, revealed that the river sediment had an average of 12.1 to 22.3 synthetic microfibers per 100 g [114]. Moreover, in some metropolitan areas, studies reported microplastics with 517,000 particles per m^2 [114]. Some researchers found that there is a positive and significant correlation between population density and microplastic concentrations [106].

The pervasive reach of microplastics extends across diverse ecosystems, with the documented accumulations in beaches, coastal sediments, arctic snow, Antarctic ice, surface, and groundwater, as well as the deep expanses of seas and oceans. Land emerges as a primary contributor, releasing an estimated 12.5 million tons of plastic annually, becoming a major source of marine pollution [115]. Research estimates approximately 250,000 tons of microplastics adrift in the oceans, with concentrations varying based on proximity to industrial sources, as elucidated by Talbot and Chang [116]. Notably, plastic particle concentrations tend to decrease gradually with distance from these sources, though concentrations surge in urbanized areas with dense populations. This spatial variation is critical, as in these densely populated regions, humans are more likely to encounter microplastics, potentially through the consumption of contaminated food and water, as highlighted by Kwon et al. [117]. This intricate interplay of environmental distribution underscores the need for a comprehensive understanding of the dynamics driving microplastic pollution and its potential implications for human exposure and ecological health.

3.5 Microplastic on beaches and sediment

Microplastics are pervasive micropollutants in marine environments, including beaches and sediments. These tiny particles originate from various sources, such as the breakdown of larger plastic debris, microbeads in personal care products, and synthetic fibers from clothing [118]. Recent studies demonstrated that microplastics are present in large quantities on beaches worldwide, with varying concentrations based on local human activities and ocean currents [119]. For example, a 2024 study conducted on the beaches of Badung, Bali, found an average of 90.7 particles/kg sediment, with Kuta Beach exhibiting the highest concentration due to its popularity among tourists [119].

The presence of microplastics in the marine sediments poses a real threat to the benthic-dwelling organisms and overall ecosystem health. Microplastics can be ingested by marine biota, leading to physical blockages, reduced feeding, and potential exposure to toxic chemicals associated with plastics [120]. Recent research studies have highlighted the widespread distribution of microplastics in sediments, even in remote areas, indicating their persistence and ability to be transported over long distances in marine ecosystems [120]. In a recent 2024 review, the authors emphasized the need for standardized extraction techniques to accurately assess their abundance and impact [118]. Additionally, a research study on sediment interactions in estuarine environments demonstrated how microplastics reacts and behaves under different conditions, further underscoring their complex dynamics [120].

Efforts to mitigate microplastic pollution in marine environments (both sediment and water) are ongoing, with a special focus on reducing plastic waste at the source and improving waste management practices. Innovative methods for extracting microplastics from sediments, such as density flotation techniques, have evolved to enhance the efficiency and accuracy of microplastic quantification [118]. Moreover, public awareness campaigns and policy measures aimed at reducing single-use plastics and promoting sustainable alternatives are crucial in addressing the major causes of microplastic pollution [119]. Continued research and collaboration are essential to develop effective strategies for managing and mitigating the impacts of microplastics on the marine environment [120].

4. Sampling strategies for microplastics

Sampling microplastics demands a meticulous approach due to their diminutive size and pervasive dispersion. Beginning by strategically selecting sites, potential sources like urban or industrial areas are considered, opting for various environmental media such as water, sediment, or biota. Microplastics sampling employs a variety of techniques, each with its strengths and limitations that must be considered (**Table 3**). Specialized equipment, like fine-mesh plankton nets or sediment cores, are employed and depth profiling is implemented to grasp the vertical distribution of microplastics. Nets are the most common method used for sampling microplastics due to their capability of collecting large volumes of water [1]. There are various types of nets including neuston net, plankton net, manta net, continuous net, and manual net [9]. Polyvinyl chloride (PVC) is commonly used for constructing nets. Pumping systems could be used for sampling microplastics in water columns and could work for several hours either at the same transect or over different sampling stations [121]. In addition to these common techniques, there are several other emerging microplastics sampling

Sampling techniques	Remarks	Strengths	Limitations
Net sampling	Net sampling involves using a net to filter microplastics out of water or sediment samples. Net sampling is relatively inexpensive and easy to conduct, but it is not very efficient at capturing small microplastics (less than 100 micrometers in size)	Inexpensive Easy to conduct	Not very efficient at capturing small microplastics
Pump sampling	Pump sampling involves using a pump to filter microplastics out of water samples. Pump sampling is more efficient at capturing small microplastics than net sampling, but it is also more expensive and time-consuming to conduct.	Efficient at capturing small microplastics	Expensive Time-consuming
Grab sampling	Grab sampling involves collecting a sample of water or sediment by hand. Grab sampling is simple and inexpensive, but it is not very representative of the entire sampling area.	Simple Inexpensive	Not very representative of the entire sampling area
Core sampling	Core sampling involves collecting a cylindrical sample of sediment using corer. Core sampling can provide a detailed record of microplastic accumulation over time, but it is more expensive and time-consuming to conduct than other sampling techniques.	Provides a detailed record of microplastic accumulation over time	Expensive Time-consuming

Table 3.
Different microplastics sampling techniques, each with its strengths and limitations.

techniques, such as passive sampling and drone-based sampling. These techniques offer several advantages over traditional techniques, but they are also more expensive and time-consuming to conduct. Quantitative sampling is ensured through systematic or random approaches, minimizing contamination risks by using clean equipment and proper protective gear. Samples are preserved in uncontaminated containers and size fractionation is considered to delve into specific particle ranges. Biological samples are integrated to comprehend bioaccumulation and food web dynamics. Passive sampling techniques are explored for temporal insights, accounting for seasonal variations. Data collection methods are standardized for cross-study comparisons, and foster collaboration with researchers and citizen scientists to enhance spatial coverage. Tailoring your strategy to study goals and microplastic characteristics is paramount for a successful investigation. Microplastic sample preparation, separation, purification, and digestion can be found in detail in these studies [99].

4.1 Microplastic polymer characterization

Identifying microplastics would require polymer identification, which involves advanced techniques to identify their chemical composition, size, shape, and surface properties. Spectroscopic techniques such as Fourier-transform infrared spectroscopy (FTIR) and Raman spectroscopy are widely used for identifying the chemical composition of microplastics [122]. FTIR measures the absorption of infrared light by the sample, providing a molecular fingerprint that can identify different types of polymers. On the other hand, Raman spectroscopy uses the scattering of light to provide information about molecular vibrations, which can also identify polymer types. These techniques are non-destructive and suitable for small particles, making them ideal for microplastic analysis (**Table 3**).

Microscopy techniques such as scanning electron microscopy (SEM) and transmission electron microscopy (TEM) are used to examine the morphology and surface characteristics of microplastics [128]. SEM provides detailed images of the surface structure, while TEM offers insights into the internal structure of the microplastic particles. These methods are often combined with energy-dispersive X-ray spectroscopy (EDS) for elemental analysis. Additionally, thermal analysis techniques such as differential scanning calorimetry (DSC) and thermogravimetric analysis (TGA) study the thermal properties of microplastics. DSC measures heat flow associated with phase transitions, helping identify polymers based on their melting points, while TGA measures weight loss during heating, providing information on thermal stability and composition.

5. Effect of microplastics

Plastics play a pivotal role in various sectors, serving purposes ranging from food packing, drug delivery, and fuel to safety, insulation, and one-use medical kits [129]. However, their versatile applications come with a stark downside—plastics pose

Potential toxic effect	Description	References
Acute toxicity	Ingestion of MPs may lead to physical blockages in the digestive tract, causing pain, nausea, and vomiting.	[123–127]
(Sub)chronic toxicity	Repeated exposure to MPs may lead to long-term health effects, including inflammation, oxidative stress, and genotoxicity.	
Carcinogenicity	Some studies suggest MPs may act as carcinogens, increasing the risk of developing cancer.	
Cytotoxicity	MPs may damage cells directly, leading to tissue damage and organ dysfunction.	
Neurotoxicity	MPs may disrupt the nervous system, leading to neurological problems.	
Immune system disruption	MPs may suppress the immune system, making it harder to fight off infections and diseases.	
Transfer to other tissues	MPs may be transported through the bloodstream and lymphatic system, reaching other organs and tissues where they can cause further damage.	
Endocrine system disruption	Some MPs may mimic hormones and disrupt the endocrine system, leading to reproductive problems and other health issues.	
Developmental toxicity	Exposure to MPs during pregnancy may harm the developing fetus.	

Table 4.
Some of the possibly toxic impacts of MPs on human health.

significant harm to human health (**Table 4**) [129] and contribute to elevated pollution levels in air, land, and water [130].

5.1 Terrestrial ecosystem

Microplastics make their way into agricultural soils through various pathways like irrigation water and sewage sludge application as fertilizer, littering, and atmospheric fallout [131]. Once in the soil, they can persist for decades and interact with terrestrial biota and ecological processes. Exposure studies reveal microplastics can reduce seed germination rates and plant growth [132]. Earthworms and microarthropods ingest the particles, which reduces their populations and impairs the soil function that these organisms support [133]. **Figure 3** illustrates the percentage of MPs focused on various terrestrial organism groups, highlighting the effects of MPs accumulation on these groups. These groups are highly impacted due to their ecological and commercial significance. Terrestrial organisms are increasingly exposed to microplastics (MPs) through various pathways, with mammals showing the highest exposure, primarily from contaminated soil, water, and food. MPs cause inflammation, oxidative stress, and endocrine disruption [134]. Birds ingest MPs *via* seeds, insects, and polluted habitats, leading to gastrointestinal blockages and bioaccumulation of toxins [135]. Reptiles encounter MPs through soil, water, or prey, with potential impacts on thermoregulation and prey availability [136]. Amphibians face risks in polluted freshwater habitats, including the developmental disruption and genotoxicity [137]. These findings underline the varied ecological risks of MPs across terrestrial taxa. Microplastic exposure may also trigger effects in the gut microbiomes of organisms, even without evidence of ingestion. Such exposure has been found to alter the gut microflora, affect isotopic signature, and display deleterious effects on the growth and reproduction of springtails [138]. **Table 5** displays some studies investigating the influences of MPs on aquatic organisms.

At the subsurface, microplastics are incorporated into soil aggregates and pore spaces, leading to changes in bulk density, water retention, and the stability of aggregates that preserve soil carbon stocks [138]. The accumulation of microplastics in natural and agricultural soils raises concerns over food safety as well as impacts on belowground ecosystems underlying terrestrial environments (**Figure 4**).

The improper disposal of plastics in landfills and inadequate recycling practices give rise to the release of toxic elements, such as lead and cadmium [150]. This not only exacerbates environmental pollution but also has severe repercussions on human health, affecting reproductive success, inducing oxidative stress, influencing metabolic behavior, hampering growth, and contributing to histopathological issues [129].

There are also physical effects that can arise from microplastic contamination. Larger plastics can limit the exchange of gases and compounds, which may affect environmental health and cause organism entanglement. Smaller particles can be ingested or inhaled, which may cause pseudo-satiation, blockage of the digestive tract, or abrasion and irritation of mucosa [138].

5.2 Aquatic ecosystems

In oceans, lakes, and rivers globally, aquatic life across trophic levels ingests microplastics through direct or indirect pathways. Studies have detected microplastic particles in the digestive tracts of marine mammals, commercial fish, shellfish, and seabirds that mistake them for prey [151]. Microplastics have a significant impact on

Category	Effects	Remarks	References
Plants	Reduced seed germination and plant growth	Exposure to Mps led to decreased seed germination and inhibited root elongation in wheat plants.	[133]
	Altered nutrient uptake	Mps can interfere with nutrient uptake in plants, affecting their overall growth and development.	[139]
	Disrupted root development	Mps can disrupt root development in plants, leading to reduced biomass and impaired nutrient acquisition.	[140]
	Increased oxidative stress	Mps can induce oxidative stress in plants, which can negatively impact their physiological processes in cucumber (*Cucumis sativus* L.).	[141, 142]
Invertebrates	Ingestion of microplastics leads to physical blockage of digestive systems	Ingestion of MPs by earthworms can lead to physical blockages in their digestive tracts, affecting their feeding and nutrient absorption	[143]
	Reduced feeding and growth rates	Exposure to microplastics can reduce feeding rates and growth in soil-dwelling invertebrates, such as springtails.	[144]
	Altered behavior and reproduction	Exposure to MPs can alter the behavior and reproduction of soil-dwelling invertebrates such as *Caenorhabditis elegans*, affecting their population dynamics.	[145]
	Accumulation of toxic chemicals associated with microplastics	MPs can act as carriers for toxic chemicals, which can accumulate in the tissues of invertebrates upon ingestion.	[52]
Vertebrates	Ingestion of microplastics leading to internal injuries and blockages	MPs can cause internal injuries and blockages in the digestive systems of birds, leading to reduced feeding efficiency and overall health	[146]
Soil microbes	Impact soil microbial communities and enzymatic activities	MPs can impact soil microbial communities and enzymatic activities. The presence of MPs fosters the growth of bacteria and fungi, altering the properties of MP materials. This can affect their bioavailability, degradability, and mobility ultimately posing a threat to crops connected to the soil's microbiome and fauna.	[20, 147]
	Reduce respiration	It has been noted that all types of MPs, except fragments, show a decreasing impact on cellobiosidase activity.	[148]

Table 5.
Some studies related to the influences of MPs on terrestrial biota.

aquatic organisms, as illustrated in **Figures 5** and **6**, which highlight the percentage of MPs according to studies conducted on different groups of aquatic organisms [149]. Fish represent the most extensively studied group. This is primarily due to their role in human consumption and their susceptibility to microplastic ingestion, which causes gastrointestinal blockages, reduced feeding efficiency, and bioaccumulation of toxic

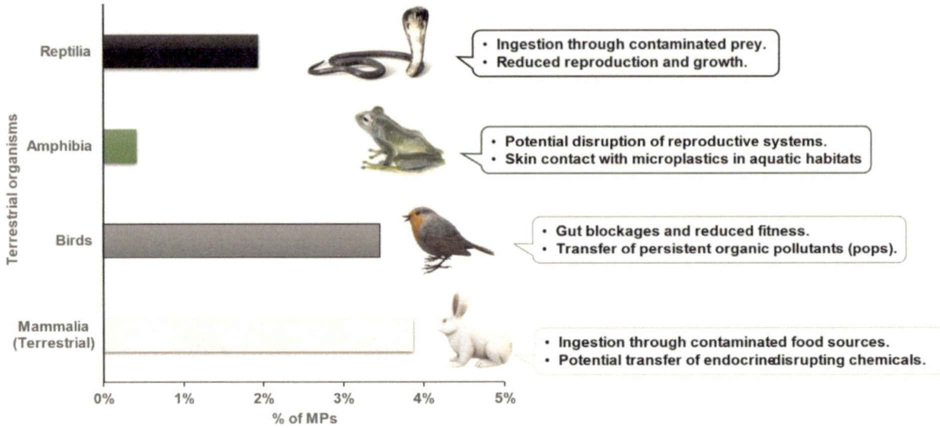

Figure 4.
Percentage of microplastic and their impacts across various terrestrial organism groups. The chart highlights the differences in research focus on aquatic organisms [149].

substances [152]. Small and large crustaceans, collectively representing 20% of the studies, are also highly affected, showing decreased survival rates and impaired molting due to microplastic ingestion [153]. Mollusks, accounting for 10% of the research, exhibit reduced filtration efficiency and oxidative stress, particularly in species such as mussels and oysters [154].

Other aquatic groups such as annelids, marine mammals, echinoderms, cnidarians, rotifers, and poriferans are less studied but still show diverse effects, including bioaccumulation, impaired reproduction, and disruptions to ecosystem functions [149]. These impacts highlight the need for further studies on underrepresented groups to gain a comprehensive understanding of microplastic pollution in aquatic environments. Primary producers such as algae and sea grasses exposed to microplastics register effects on photosynthesis and growth that can cascade up food webs (**Table 6**) [163].

Beyond physical harm from intestinal blockage and reduced nutritional intake, microplastics also leach additive chemicals as they break down and serve as carriers for other waterborne hydrophobic pollutants [11]. Microplastics reach the food chain through adhering and/or absorbing hydrophobic traces such as polybrominated diphenyl ethers (PBDEs), polychlorinated biphenyls (PCBs), and polycyclic chlorinated pesticides (PCPs). Such traces are the main sources of hazardous pollutants and get into the food web *via* aquatic biota [164]. These combined exposures elicit toxic responses including inflammation, endocrine disruption, neurological dysfunction, and population-level impacts through reproductive failure seen in species studied *in situ* [165]. The mixture of plastic waste with legacy contaminants poses ecological risks that demand closer examination.

5.3 Atmospheric

While research on airborne microplastics is still evolving, evidence confirms the presence of fibers, fragments, and particles in the atmosphere. Sources range from synthetic textile fibers to erosion of terrestrial plastic pollution and sea spray. Atmospheric transport enables microplastics to reach remote regions such as mountains and the Arctic [166]. Variable factors such as wind patterns, turbulence, and proximity to

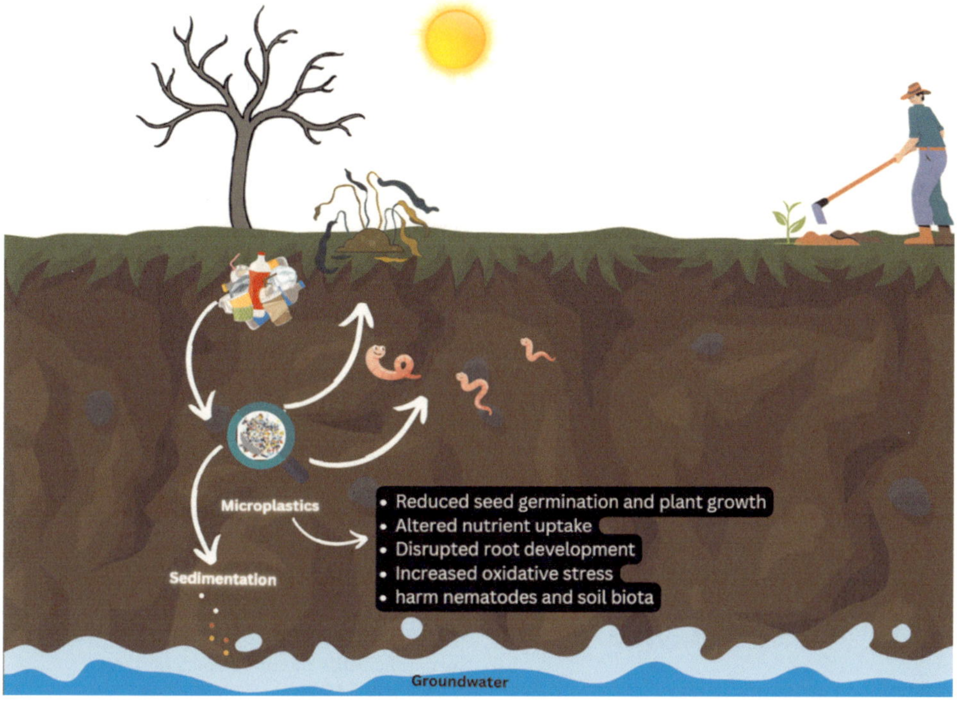

Figure 5.
A visual representation illustrating the different ways microplastics (MPs) might impact the terrestrial ecosystem.

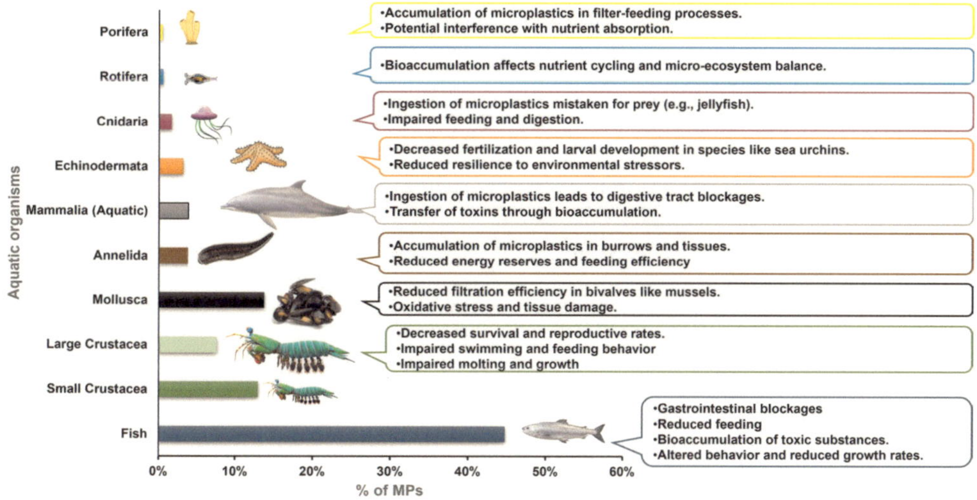

Figure 6.
Percentage of microplastic and their impacts across various aquatic organism groups. The chart highlights the differences in research focus on aquatic organisms [149].

point sources influence concentrations settling on land and water bodies through dry and wet deposition [167].

Inhalation studies associate airborne microplastics with respiratory irritation and inflammation [168]. Beyond direct health effects, atmospheric circulation and

Category	Effects	Remarks	References
Phytoplankton	Photosynthetic, electron transport and growth inhibition; oxidative stress	Low-density polyethylene MPs cause a decline in photosynthetic activity, oxidative stress, and metabolism alternation of *Chaetoceros calcitrans* microalgae	[5]
		Diminished algal cell viability, photosynthetic, electron transport, rise in oxidative stress in fresh water algae, *Scenedesmus obliquus,* and *Chlorella* sp.	[155, 156]
Zooplankton	Histopathology; inflammation; oxidative stress; microbiome disruption; DNA damage; gill abnormalities	The consumption of MPs showed damage to the intestines, diminished intake, hindered growth, restricted reproductive output, shortened lifespan, and abnormal or fatal gene expression such as *Daphnia magna*	[157, 158]
Aquatic vertebrates	Reduced feeding efficiency and nutrient absorption	Microplastic ingestion can impair the feeding efficiency and nutrient absorption in reptiles, affecting their growth and reproductive success.	[159]
	Altered behavior and reproduction	That exposure to microplastics can disrupt the behavior and reproduction of marine crustaceans, which can have cascading effects on their populations	[160]
	Accumulation of toxic chemicals associated with microplastics	Microplastics can accumulate and transport toxic chemicals, such as polychlorinated biphenyls (PCBs), in the tissues of fish and other vertebrates.	[161]
	Severe damage	Ingested MPs cause severe damage in *Ardenna carneipes* seabirds such as decline in tubular glands; inflammation; fibrosis; organ damage	[162]

Table 6.
Some studies related to the influences of MPs on aquatic biota.

downwind hotspots serve as key microplastic entry points into terrestrial and aquatic food chains with implications for environmental and human exposure [167]. Establishing standardized air monitoring and fate models is an urgent research need in this rapidly growing area.

5.4 Human health

Limited clinical evidence directly links microplastics to specific human health outcomes. However, cell culture, rodent, and aquatic animal studies reveal various toxicity mechanisms that provide early warning on potential effects relevant to human physiology. Ingested microplastics appear to translocate across intestinal epithelial layers into circulation and tissues [169].

Once internalized, particles generate reactive oxygen species that trigger inflammatory pathways and interfere with metabolism, gut-brain signaling, and immune function [170]. Toxic chemical additives leaching from plastics into biological fluids compound these stressors. Inhalation represents another central exposure concern based mainly on occupational studies of plastic factory workers [171]. Respiratory exposures are associated with declined lung function and airway irritation over time. Combined with ubiquitous environmental presence, plausible biological impacts warrant proactive steps to elucidate risks and inform public health protections.

5.5 Microplastic interaction with plants and animals

Microplastics interact with plants and animals in various detrimental ways. In plants, microplastics can induce stress responses, including reductions in seed germination, root elongation, and photosynthetic activity [172]. These tiny particles can alter soil properties, affecting water retention and nutrient availability, which in turn impacts plant growth, health, and overall productivity [172]. Research studies have shown that microplastics can be taken up by plant roots and translocated to other parts of the plant, potentially entering the food chain [172]. This uptake can lead to physical blockages and chemical toxicity, further stressing the plants and reducing their overall vitality [172].

In animals and through ingestion, microplastics are ingested either directly or indirectly through the food chain, leading to various health issues. For example, marine organisms such as zooplankton ingest microplastics, which can cause physical blockages and reduce their ability to feed on algae [173]. This disruption in feeding behavior can lead to decreased growth and reproductive success [173]. Additionally, microplastics can leach harmful chemicals, such as additives and pollutants, which can accumulate in the tissues of animals and cause toxic effects [173]. Research has shown that microplastics can cause physical damage to the digestive systems of marine animals, leading to inflammation, reduced nutrient absorption, and even death [173] in severe cases.

6. Controlling of MPs and strategies

Controlling microplastics in our environment poses a significant challenge due to their small size and widespread distribution. Several methods have been explored to face this problem as follows.

6.1 Elimination, reduction, and reuse

Many national and local authorities proposed laws and regulations to be implemented to reduce the plastic discharge into the environment. However, such regulations are either not in action or ineffective in removal. Many countries, including China, Europe, and the USA, have prohibited throwaway plastics—known as single-use plastics—since 2008. However, there are hundreds of supermarkets and shopping malls that are still using plastic bags. In the European Union and as a part of the green economy transition plan, the use of disposable plastic bags will be significantly minimized and, by 2030, should be replaced by recyclable packing [101]. Moreover, the use of microbeads is prohibited in several nations since they are the primary source of microplastics in the aquatic environment [8]. Reducing plastic fibers should be implemented by optimizing the optimal washing schemes. Ultimately, reducing plastic usage and replacing plastics with paper bags and/or other recyclable materials would be the most effective way to prevent microplastics from reaching the food web chain.

It has been estimated that only 9% of the over 8.3 billion metric tons of plastic produced globally has been recycled [174]. There are substantial opportunities to improve recycling, especially for complex plastic components in electronic devices and automobiles [175]. The technology exists to recycle and repurpose thermoplastics from electronics and shred and reform thermoset materials from automobiles [176]. For example, the Eastman Tritan Renew copolyester introduced by Eastman Chemical Company uses a chemical recycling process allowing for infinite recyclability of plastics (Eastman Chemical Company, TN, USA). Improving collection infrastructure and implementing standardized labeling for plastic types to support sorting for recycling have also been identified as impactful solutions [177].

6.2 Removal using filtration and sorption methods

Due to microplastic bioaccumulation and low degradability in marine environments, the removal techniques should be a top priority. They have a high sorbent capacity for waterborne contaminants and other pollutants, facilitating the translocation of pollutants into new habitats [178]. Some researchers investigated the adherence capacity of microplastic particles on the surface of microorganisms including algae [179], seaweeds [180], and bacteria [181]. For example, an experimental study [180] conducted on a seaweed *Fucus vesiculosus* revealed that the removal efficiency of microplastic was improved by 94.5%. The authors further indicated that microplastic adsorption on the seaweed's surface was a result of alginate (anionic polysaccharide substance) release from the cell wall, which has gelatinous characteristics that played a critical role in the adherence of the polystyrene particles to the algal surface—improving the removal efficiency [182]. Another study conducted by Nolte et al. on the removal of 20–500-nm polystyrene materials using unicellular algae, *Pseudokirchneriella subcapitata*, revealed higher adsorption and removal efficiency of microplastic particles of polystyrene material [183]. Notably, high concentrations of these microplastics may have deleterious effects on such microorganisms. The detrimental effects are evident, as highlighted by the significant physical damage observed in marine microalgae *Chaetoceros calcitrans* cell structure when exposed to high concentrations of low-density polyethylene microplastics. This underscores the need for increased attention to the ecological footprint of microplastics, especially considering the ongoing breakdown of plastics within the ecosystem. Recognizing and addressing these effects on microorganisms are crucial steps in mitigating the broader environmental consequences of microplastic pollution [5].

Membrane technology is another emerging research area where scientists developed a dynamic membrane that showed promising results for microplastic particle removal from wastewater [184]. Some researchers concluded that wastewater turbidity was significantly decreased from 195 to 1 NTU *via* microplastic particle removal [185]. In large-scale applications, membrane bioreactors are in use for removal of large capacities of microplastics [186]. Bioreactor applications might be of a high-cost concern and depend largely on membrane durability, wastewater influent flux rate, and the contamination levels of microplastics [187].

6.3 Plastic biodegradation

To show the impressive role of microorganisms (both bacteria and archaea) and their role in microplastic removal through biodegradation in coastal sediment, Harrison et al. [188] highlighted several research areas and gaps where scientists need to focus more to advance our understanding of microplastic biodegradation through interaction with microorganisms. Several research studies have reported biodegradation of both synthetic and naturally occurring microplastics *via* microorganisms' interactions [189]. Ongoing research on the role and evidence of zooplankton in microplastic degradation is still evolving. Shrinking in size of plastic particles that were viewed under an electron microscope was clear evidence of the biotransformation of microplastics to nanoplastics in biologically mediated processes [190]. Many other research studies were conducted to investigate the biological degradation of microplastics *via* marine fungi [191] and some crustaceans [192]. For example, *Ideonella sakaiensis* 201-F bacterium produces the enzymes PETase and MHETase, which use hydrolysis to break down the plastic polyethylene terephthalate (PET). These enzymes work best at temperatures around 70–75° C [193], beyond this specific bacterium, LC-cutinase from compost-dwelling bacteria, extremophilic bacteria tolerant of very high/low salt/pH levels, and common plastic-degrading species such as *Pseudomonas* and *Bacillus* [194], while *Aspergillus oryzae* fungus secretes hydrolysis enzymes that hydrolyze the plastic polymer polybutylene succinate (PBS) optimally at 50°C [195]. Through enzyme secretion, colonization, and other mechanisms, these diverse microorganisms can fragment plastic products such as PVC, PP, PE, and PC, facilitating biodegradation under proper conditions like the presence of co-substrates and appropriate temperature ranges.

6.4 Enhancing microplastic removal efficiency

Wastewater treatment plants are one of the major sources of microplastics in the local environment. Accordingly, improving the removal efficiency of plastics at those facilities will tremendously reduce the microplastic concentration in the aquatic environment [102]. Increasing the clearance efficiency of plastics at wastewater treatment facilities could reach 98% [196] *via* methods like electrocoagulation, where microplastic removal could reach 99.24% [197], especially through membrane bioreactor applications.

There are other chemical techniques for improving microplastic removal, including coagulation/flocculation [198], chlorination [199], agglomeration [200], reduced graphene oxides [201], and nanoparticle films [202]. For example, the removal efficiency of microplastics *via* coagulation could reach 98.2% by poly-aluminum chloride [26]. On the other hand, biological techniques for microplastic removal are less efficient than chemical removal, but biological strategies have low operating costs. Among biological techniques for microplastic removal are activated sludge [203], biofilter [204], active ingestion [205], and biodegradation [206].

7. Wetland ecosystems and their role in microplastic removal

Wetlands are unique ecosystems that provide a wide variety of ecosystem services and functions, including soil carbon regulation [207–212]. However, many research studies showed evidence of microplastic accumulation in wetland sediment and biota [53, 150]—which in turn have adverse effects on ecosystem services offered by wetland habitats.

Assessment of microplastic accumulation, extraction, identification, and their impacts on wetland habitats received less attention from researchers. Kumar et al. [53] identified certain limitations and research shortcomings in microplastic research in wetlands, including sources, pathways, sampling approaches, adsorption mechanisms, policy intervention, and risk assessment. Moreover, the concentration of microplastics in surface water [213] has minimal data available to researchers, which is considered a limitation for modeling the fate and transport of microplastics [214, 215] especially in wetland ecosystems. Moreover, microplastic research studies in other wetlands such as meadows, seasonal wetlands, and constructed wetlands are very rare and limited.

Wetland vegetation plays a crucial role in microplastic removal since they are working as a natural filter [216], and vegetation incorporation in newly constructed wetlands is recommended for microplastic removal. Moreover, clay particle size in wetlands impacts the microplastic abundance, where higher microplastic abundance was associated with finer clay [217]. Both natural and constructed wetlands have a fundamental role in removing suspended particles (90–97% removal efficiency) from the water column [218]. Constructed wetlands are commonly used for wastewater treatment [219]; however, the fate and transport of microplastics in the newly constructed and treated wetlands are poorly understood [220].

Different wetland plant species have an additional capacity to accumulate various pollutants [16, 107]. It is highly recommended when constructing wetlands for remediation to avoid non-native plant species [221] as they can become invasive species that cause tremendous environmental problems including lower native biodiversity [107, 222–224]. Some wetland ecosystems such as mangroves and coral reefs are susceptible to plastic contamination [225, 226], and protecting them should be a top priority. The detection of microplastics in different wetlands including mangroves [226], marine environment [227, 228], and food chain of different biota including Nile Tilapia [41, 229] presents an alarming signal of increasing plastic loads to the aquatic environments. In conclusion, the science required for the assessment of emerging contaminants of microplastics in wetlands is necessary for identifying the adverse impacts on wetland ecosystem services and functions.

8. Key challenges in controlling microplastic (MPs) pollution

The heterogeneous physicochemical properties of MPs pose a significant challenge for large-scale surveillance, risk assessment, and remediation. Their ecotoxicity effects remain poorly characterized [230]. Furthermore, public perceptions, attitudes, and behaviors toward MPs are underexplored, hindering efforts to motivate consumer and policy actions [231]. Additional major hurdles include the lack of reliable analytical techniques for proper MP identification and quantification in complex environmental matrices as well as in food [232]. Implementing policies to curb plastic usage also faces barriers once interventions are enacted, demanding improved governance. Moreover, the current lack of standardized procedures for collecting and analyzing environmental MP data limits comparative assessments, especially for emerging pollutants such as nanoplastics. Similarly, the absence of a comprehensive spectral database of polymer types constrains precise analysis [233].

The interaction of microplastics with environmental matrices alters their infrared spectra, complicating polymer identification and modification studies [93]. Moreover, plastic recycling faces issues due to the mixing of polymer types and additives that reduce purity and make the extraction of desired materials complex [234]. The highly stable carbon-carbon backbone of polymers like polyethylene also impedes biodegradation. Additional analytical barriers include collecting adequate MP quantities from complex environmental samples for chemical analysis, as well as the low detection frequencies and high limits posed by tiny particle sizes [235].

In summary, microplastic variability with environmental weathering, issues with plastic mixtures in recycling, and analytical chemistry challenges around detecting low abundance MP analytes in complex matrices present further roadblocks to research and remediation efforts. Multi-residual analytical tools and polymer extraction methods may help resolve some constraints [235]. Ultimately, the extremely stable chemical structure of many plastic polymers continues to pose an overarching challenge for developing mitigation strategies.

9. Future directions

Microplastics pollution is among emerging contaminants that have gathered the attention of the scientific community in the last two decades [26]. The research focus was mainly on microplastic sources, sinks, and distribution in the environment including land, rivers, lakes, and aquatic ecosystems. Limited research studies were conducted on microplastics removal efficiency and fundamental research gaps about this aspect are still poorly understood. Moreover, the fate and transport mechanisms of plastics are still considered challenging research aspects that require state-of-the-art investigation. Moreover, ecological risk assessment and ecotoxicological studies that investigate the microplastic impacts on microorganisms, humans, and animals have very limited data and remain unexplored aspects that require in-depth research studies. Improving and increasing microplastics removal efficiency should be a top-priority research experiment.

Previous research studies extensively focused on the role of microalgae and other microorganisms in improving the removal efficiency of microplastic particles from marine environments. They presented the higher capacity of algae in removing microplastic particles, especially with those of positive charges, which in turn increased the waterway's quality bioindicators. Membrane bioreactors are very convenient for microplastic removal from wastewater runoff on large scales, especially for giant wastewater treatment facilities but might be of a high-cost concern. Aside from the mechanical and production of synthetic products used for microplastic removal, biodegradable processes mediated by microorganisms are available. Microorganisms that were involved in plastic degradation include bacteria, archaea, fungi, and some crustaceans.

Accordingly, studies on the role of the constructed and treated wetlands showed promising results for microplastic removals and in-depth research should continue for following: (1) a better understanding of microplastic fate and transport in the wetland ecosystems; and (2) developing cost-effective and sustained remediation technologies for microplastic removal. It is highly recommended that an establishment of consensus-based guidelines for microplastic threshold concentrations be established for better implementation of removal efficiency techniques.

10. Conclusions

Microplastics are insoluble in water and hard to see by the naked eye. Based on the origin, they can be classified into primary and secondary. They exist in the air, land,

and aquatic environments where they enter the food web. They are very harmful to human health and aquatic biota. Improving and increasing the microplastic removal efficiency is highly recommended to reduce the microplastic loads to the aquatic environment. There are different ways and techniques of microplastic removal that are still ongoing and require more development to increase removal efficiency. These techniques include dynamic membranes, bioreactor membranes, and biological removal. Ingestion of microplastics by organisms is still under investigation to be considered a removal strategy. Despite its high removal efficiency, adsorption removal techniques still have a secondary pollution risk. On the other hand, biodegradation of microplastic mediated by biological processes poses no risks in the environment but usually has low degradation efficiency. Local authorities and governmental agencies should regulate plastic usage. Not only should minimizing plastic use be the goal of the government, but also public awareness should be raised for replacing plastics with recyclable items like paper bags at grocery stores and shopping malls. Moreover, plastic discharge into wastewater should be strictly prohibited, especially since most wastewater treatment facilities cannot fully remove microplastic. Pollution assessment for microplastics should be strengthened in lands, air, and aquatic environments to provide holistic insights into the level of plastic distribution mechanisms, sinks, and sources.

Acknowledgements

The first author expressed his deep gratitude to STEM undergraduate students (A.G., M.S., and P.H.) who conducted the initial literature assessment.

Author details

Amr E. Keshta[1,2*], Alaa Gamal[1], Melad Soryal[1], Peter Hana[1], Bader Z. Albogami[3] and Mostafa Elshobary[1,4]

1 Botany and Microbiology Department, Faculty of Science, Tanta University, Tanta, Egypt

2 Smithsonian Environmental Research Center, Edgewater, Maryland, USA

3 Department of Biology, Faculty of Arts and Sciences, Najran University, Najran, Saudi Arabia

4 Aquaculture Research, Alfred Wegener Institute (AWI) – Helmholtz Centre for Polar and Marine Research, Am Handelshafen, Bremerhaven, Germany

*Address all correspondence to: akeshta@umd.edu

References

[1] Cutroneo L, Reboa A, Besio G, Borgogno F, Canesi L, Canuto S, et al. Microplastics in seawater: Sampling strategies, laboratory methodologies, and identification techniques applied to port environment. Environmental Science and Pollution Research. 2020; **27**(9):8938-8952. DOI: 10.1007/s11356-020-07783-8

[2] Geyer R, Jambeck JR, Law KL. Production, use, and fate of all plastics ever made. Science Advances. 2017;**3**(7): e1700782. DOI: 10.1126/sciadv.1700782

[3] Scalenghe R. Resource or waste? A perspective of plastics degradation in soil with a focus on end-of-life options. Heliyon. 2018;**4**(12). DOI: 10.1016/j.heliyon.2018.e00941

[4] Zheng XW, Liu XL, Zhang LL, Wang ZM, Yuan Y, Li J, et al. Toxicity mechanism of nylon microplastics on *Microcystis aeruginosa* through three pathways: Photosynthesis, oxidative stress and energy metabolism. Journal of Hazardous Materials. 2022;**426**. DOI: 10.1016/j.jhazmat.2021.128094

[5] Senousy HH, Khairy HM, El-Sayed HS, Sallam ER, El-Sheikh MA, Elshobary ME. Interactive adverse effects of low-density polyethylene microplastics on marine microalga Chaetoceros calcitrans. Chemosphere. 2023;**311**. DOI: 10.1016/j.chemosphere.2022.137182

[6] Primpke S, Wirth M, Lorenz C, Gerdts G. Reference database design for the automated analysis of microplastic samples based on Fourier transform infrared (FTIR) spectroscopy. Analytical and Bioanalytical Chemistry. 2018; **410**(21):5131-5141. DOI: 10.1007/s00216-018-1156-x

[7] Andrade MC, Winemiller KO, Barbosa PS, Fortunati A, Chelazzi D, Cincinelli A, et al. First account of plastic pollution impacting freshwater fishes in the Amazon: Ingestion of plastic debris by piranhas and other serrasalmids with diverse feeding habits. Environmental Pollution. 2019;**244**:766-773. DOI: 10.1016/j.envpol.2018.10.088

[8] Rostami S, Talaie MR, Talaiekhozani A, Sillanpaa M. Evaluation of the available strategies to control the emission of microplastics into the aquatic environment. Environmental Science and Pollution Research. 2021;**28**(15):18908-18917. DOI: 10.1007/s11356-021-12888-9

[9] Zhang SL, Wang JQ, Liu X, Qu FJ, Wang XS, Wang XR, et al. Microplastics in the environment: A review of analytical methods, distribution, and biological effects. Trac-Trends in Analytical Chemistry. 2019;**111**:62-72. DOI: 10.1016/j.trac.2018.12.002

[10] Verschoor A. Towards a Definition of Microplastics: Considerations for the Specification of Physico-Chemical Properties. Ba Bilthoven, The Netherlands: National Institute of Public Health and the Environment; 2015

[11] Koelmans AA, Bakir A, Burton GA, Janssen CR. Microplastic as a vector for Chemicals in the Aquatic Environment: Critical review and model-supported reinterpretation of empirical studies. Environmental Science & Technology. 2016;**50**(7):3315-3326. DOI: 10.1021/acs.est.5b06069

[12] Mak CW, Yeung KCF, Chan KM. Acute toxic effects of polyethylene microplastic on adult zebrafish. Ecotoxicology and Environmental Safety. 2019;**182**. DOI: 10.1016/j.ecoenv.2019.109442

[13] Bowley J, Baker-Austin C, Porter A, Hartnell R, Lewis C. Oceanic hitchhikers - Assessing pathogen risks from marine microplastic. Trends in Microbiology. 2021;**29**(2):107-116. DOI: 10.1016/j.tim.2020.06.011

[14] Barcelo D, Pico Y, Alfarhan AH. Microplastics: Detection in human samples, cell line studies, and health impacts. Environmental Toxicology and Pharmacology. 2023;**101**. DOI: 10.1016/j.etap.2023.104204

[15] Lim XZ. Microplastics are everywhere - But are they harmful? Nature. 2021;**593**(7857):22-25. DOI: 10.1038/d41586-021-01143-3

[16] Hadad HR, Maine MA. Tropical and subtropical wetland plant species used for phytoremediation in treatment wetlands. In: Otte ML, editor. Wetlands for Remediation in the Tropics: Wet Ecosystems for Nature-Based Solutions, Lobato de Magalhães, T. Cham: Springer International Publishing; 2023. pp. 31-50

[17] Keshta AE, Shaltout KH, Baldwin AH, El-Din AAS. Sediment clays are trapping heavy metals in urban lakes: An indicator for severe industrial and agricultural influence on coastal wetlands at the Mediterranean coast of Egypt. Marine Pollution Bulletin. 2020; **151**:110816

[18] Wijekoon P, Koliyabandara PA, Cooray AT, Lam SS, Athapattu BCL, Vithanage M. Progress and prospects in mitigation of landfill leachate pollution: Risk, pollution potential, treatment and challenges. Journal of Hazardous Materials. 2022;**421**. DOI: 10.1016/j.jhazmat.2021.126627

[19] Yu HW, Qi WX, Cao XF, Hu JW, Li Y, Peng JF, et al. Microplastic residues in wetland ecosystems: Do they truly threaten the plant-microbe-soil system? Environment International. 2021;**156**. DOI: 10.1016/j.envint.2021.106708

[20] Fei YF, Huang SY, Zhang HB, Tong YZ, Wen DS, Xia XY, et al. Response of soil enzyme activities and bacterial communities to the accumulation of microplastics in an acid cropped soil. Science of the Total Environment. 2020;**707**. DOI: 10.1016/j.scitotenv.2019.135634

[21] Zang HD, Zhou J, Marshall MR, Chadwick DR, Wen Y, Jones DL. Microplastics in the agroecosystem: Are they an emerging threat to the plant-soil system? Soil Biology and Biochemistry. 2020;**148**. DOI: 10.1016/j.soilbio.2020.107926

[22] Pignattelli S, Broccoli A, Renzi M. Physiological responses of garden cress (*L. sativum*) to different types of microplastics. Science of the Total Environment. 2020;**727**. DOI: 10.1016/j.scitotenv.2020.138609

[23] Yu H, Fan P, Hou JH, Dang QL, Cui DY, Xi BD, et al. Inhibitory effect of microplastics on soil extracellular enzymatic activities by changing soil properties and direct adsorption: An investigation at the aggregate-fraction level. Environmental Pollution. 2020; **267**. DOI: 10.1016/j.envpol.2020.115544

[24] Qi YL, Yang XM, Pelaez AM, Lwanga EH, Beriot N, Gertsen H, et al. Macro- and micro- plastics in soil-plant system: Effects of plastic mulch film residues on wheat (*Triticum aestivum*) growth. Science of the Total Environment. 2018;**645**:1048-1056. DOI: 10.1016/j.scitotenv.2018.07.229

[25] Machado AAD, Lau CW, Till J, Kloas W, Lehmann A, Becker R, et al. Impacts of microplastics on the soil biophysical environment. Environmental Science & Technology.

2018;**52**(17):9656-9665. DOI: 10.1021/acs.est.8b02212

[26] Ahmed R, Hamid AK, Krebsbach SA, He JZ, Wang DJ. Critical review of microplastics removal from the environment. Chemosphere. 2022;**293**. DOI: 10.1016/j.chemosphere.2022.133557

[27] Auta HS, Emenike CU, Fauziah SH. Distribution and importance of microplastics in the marine environment: A review of the sources, fate, effects, and potential solutions. Environment International. 2017; **102**:165-176. DOI: 10.1016/j.envint.2017.02.013

[28] Issac MN, Kandasubramanian B. Effect of microplastics in water and aquatic systems. Environmental Science and Pollution Research. 2021;**28**(16): 19544-19562. DOI: 10.1007/s11356-021-13184-2

[29] Hasan Anik A, Hossain S, Alam M, Binte Sultan M, Hasnine MDT, Rahman MM. Microplastics pollution: A comprehensive review on the sources, fates, effects, and potential remediation. Environmental Nanotechnology, Monitoring & Management. 2021;**16**: 100530. DOI: 10.1016/j.enmm.2021.100530

[30] Castaneda RA, Avlijas S, Simard MA, Ricciardi A. Microplastic pollution in St. Lawrence river sediments. Canadian Journal of Fisheries and Aquatic Sciences. 2014;**71**(12): 1767-1771. DOI: 10.1139/cjfas-2014-0281

[31] Carrasco-Navarro V, Nuutinen A, Sorvari J, Kukkonen JVK. Toxicity of tire rubber microplastics to freshwater sediment organisms. Archives of Environmental Contamination and Toxicology. 2022;**82**(2):180-190. DOI: 10.1007/s00244-021-00905-4

[32] Kole PJ, Lohr AJ, Van Belleghem F, Ragas AMJ. Wear and tear of tyres: A stealthy source of microplastics in the environment. International Journal of Environmental Research and Public Health. 2017;**14**(10). DOI: 10.3390/ijerph14101265

[33] Salthammer T. Microplastics and their additives in the indoor environment. Angewandte Chemie-International Edition. 2022;**61**(32). DOI: 10.1002/anie.202205713

[34] Keshta AE, Riter JCA, Shaltout KH, Baldwin AH, Kearney M, Sharaf El-Din A, et al. Loss of coastal wetlands in Lake Burullus, Egypt: A GIS and remote-sensing study. Sustainability. 2022;**14**(9): 4980

[35] Eid EM, Arshad M, Alrumman SA, Al-Bakre DA, Ahmed MT, Almahasheer H, et al. Evaluation of soil organic carbon stock in coastal sabkhas under different vegetation covers. Journal of Marine Science and Engineering. 2022;**10**(9):1234. DOI: 10.3390/jmse10091234

[36] Abd El-Hamid H, Eid EM, El-Morsy MH, Osman HE, Keshta AE. Benefits of blue carbon stocks in a coastal Jazan ecosystem undergoing land use change. Wetlands. 2022;**42**:103. DOI: 10.1007/s13157-022-01597-9

[37] Keshta AE. Phytogeographical regions of Egypt: First open-source geospatial data and its applications. BMC Research Notes. 2023;**16**(1). DOI: 10.1186/s13104-023-06528-3

[38] Shaltout K, Bedair H, El-Khalafy MM, Keshta A, Halmy MWA. The link of socioeconomic importance to the conservation status of the Mediterranean endemic plants in Egypt. Nusantara Bioscience. 2023;**15**(2):189-211. DOI: 10.13057/nusbiosci/n150207

[39] Mani T, Hauk A, Walter U, Burkhardt-Holm P. Microplastics profile along the Rhine River. Scientific Reports. 2015;**5**. DOI: 10.1038/srep17988

[40] Zhang K, Gong W, Lv JZ, Xiong X, Wu CX. Accumulation of floating microplastics behind the three gorges dam. Environmental Pollution. 2015; **204**:117-123. DOI: 10.1016/j.envpol.2015.04.023

[41] Biginagwa FJ, Mayoma BS, Shashoua Y, Syberg K, Khan FR. First evidence of microplastics in the African Great Lakes: Recovery from Lake Victoria Nile perch and Nile tilapia. Journal of Great Lakes Research. 2016; **42**(1):146-149. DOI: 10.1016/j.jglr.2015.10.012

[42] Shabaka SH, Ghobashy M, Marey RS. Identification of marine microplastics in Eastern Harbor, Mediterranean coast of Egypt, using differential scanning calorimetry. Marine Pollution Bulletin. 2019; **142**:494-503. DOI: 10.1016/j.marpolbul.2019.03.062

[43] Razaviarani V, Saudagar A, Gallage S, Shrinath S, Arab G. Comprehensive investigation on microplastics from source to sink. Clean Technologies and Environmental Policy. 2024;**26**(6):1755-1782. DOI: 10.1007/s10098-024-02738-w

[44] Kalčíková G, Bundschuh M. Aquatic biofilms—Sink or source of microplastics? A critical reflection on current knowledge. Environmental Toxicology and Chemistry. 2021;**41**(4): 838-843. DOI: 10.1002/etc.5195

[45] Osman AI, Hosny M, Eltaweil AS, Omar S, Elgarahy AM, Farghali M, et al. Microplastic sources, formation, toxicity and remediation: A review. Environmental Chemistry Letters. 2023; **21**(4):2129-2169. DOI: 10.1007/s10311-023-01593-3

[46] Li K, Jia W, Xu L, Zhang M, Huang Y. The plastisphere of biodegradable and conventional microplastics from residues exhibit distinct microbial structure, network and function in plastic-mulching farmland. Journal of Hazardous Materials. 2023;**442**. DOI: 10.1016/j.jhazmat.2022.130011

[47] Expósito N, Rovira J, Sierra J, Folch J, Schuhmacher M. Microplastics levels, size, morphology and composition in marine water, sediments and sand beaches. Case study of Tarragona coast (western Mediterranean). Science of the Total Environment. 2021;**786**. DOI: 10.1016/j.scitotenv.2021.147453

[48] Law K, Moret-Ferguson S, Maximenko N, Proskurowski G, Peacock E, Hafner J, et al. Plastic accumulation in the North Atlantic subtropical gyre. Science. 2010; **329**(5996):1185-1188. DOI: 10.1126/science.1192321

[49] Chen H, Selvam S, Ting K, Gibbins C. Microplastic pollution in freshwater systems in Southeast Asia: Contamination levels, sources, and ecological impacts. Environmental Science and Pollution Research. 2021; **28**(39):54222-54237. DOI: 10.1007/s11356-021-15826-x

[50] Aves A, Revell L, Gaw S, Ruffell H, Schuddeboom A, Wotherspoon N, et al. First evidence of microplastics in Antarctic snow. The Cryosphere. 2022; **16**(6):2127-2145. DOI: 10.5194/tc-16-2127-2022

[51] Peeken I, Primpke S, Beyer B, Gütermann J, Katlein C, Krumpen T, et al. Arctic Sea ice is an important

temporal sink and means of transport for microplastic. Nature Communications. 2018;**9**. DOI: 10.1038/s41467-018-03825-5

[52] Scherer C, Brennholt N, Reifferscheid G, Wagner M. Feeding type and development drive the ingestion of microplastics by freshwater invertebrates. Scientific Reports. 2017;**7**: 9. DOI: 10.1038/s41598-017-17191-7

[53] Kumar R, Sharma P, Bandyopadhyay S. Evidence of microplastics in wetlands: Extraction and quantification in freshwater and coastal ecosystems. Journal of Water Process Engineering. 2021;**40**:13. DOI: 10.1016/j.jwpe.2021.101966

[54] Frank Y, Vorobiev E, Vorobiev D, Trifonov A, Antsiferov D, Hunter T, et al. Preliminary screening for microplastic concentrations in the surface water of the Ob and tom Rivers in Siberia, Russia. Sustainability. 2021;**13**(1). DOI: 10.3390/su13010080

[55] Amrutha K, Warrier A. The *first* report on the source-to-sink characterization of microplastic pollution from a riverine environment in tropical India. Science of the Total Environment. 2020;**739**. DOI: 10.1016/j.scitotenv.2020.140377

[56] Robin R, Karthik R, Purvaja R, Ganguly D, Anandavelu I, Mugilarasan M, et al. Holistic assessment of microplastics in various coastal environmental matrices, southwest coast of India. Science of the Total Environment. 2020;**703**. DOI: 10.1016/j.scitotenv.2019.134947

[57] Patterson J, Jeyasanta K, Sathish N, Booth A, Edward J. Profiling microplastics in the Indian edible oyster, *Magallana bilineata* collected from the Tuticorin coast, gulf of Mannar, Southeastern India. Science of the Total Environment. 2019;**691**:727-735. DOI: 10.1016/j.scitotenv.2019.07.063

[58] James K, Vasant K, Padua S, Gopinath V, Abilash K, Jeyabaskaran R, et al. An assessment of microplastics in the ecosystem and selected commercially important fishes off Kochi, south eastern Arabian Sea, India. Marine Pollution Bulletin. 2020;**154**. DOI: 10.1016/j.marpolbul.2020.111027

[59] Xiong X, Zhang K, Chen X, Shi H, Luo Z, Wu C. Sources and distribution of microplastics in China's largest inland Lake - Qinghai Lake. Environmental Pollution. 2018;**235**:899-906. DOI: 10.1016/j.envpol.2017.12.081

[60] Di M, Wang J. Microplastics in surface waters and sediments of the three gorges reservoir, China. Science of the Total Environment. 2018;**616**: 1620-1627. DOI: 10.1016/j.scitotenv.2017.10.150

[61] Jiang C, Yin L, Wen X, Du C, Wu L, Long Y, et al. Microplastics in sediment and surface water of west Dongting Lake and south Dongting Lake: Abundance, source and composition. International Journal of Environmental Research and Public Health. 2018;**15**(10):2164

[62] Su L, Cai H, Kolandhasamy P, Wu C, Rochman C, Shi H. Using the Asian clam as an indicator of microplastic pollution in freshwater ecosystems. Environmental Pollution. 2018;**234**: 347-355. DOI: 10.1016/j.envpol.2017.11.075

[63] Bakir A, Desender M, Wilkinson T, Van Hoytema N, Amos R, Airahui S, et al. Occurrence and abundance of *meso* and microplastics in sediment, surface waters, and marine biota from the South Pacific region. Marine Pollution Bulletin.

2020;**160**. DOI: 10.1016/j.marpolbul.2020.111572

[64] Ory N, Sobral P, Ferreira J, Thiel M. Amberstripe scad *Decapterus muroadsi* (Carangidae) fish ingest blue microplastics resembling their copepod prey along the coast of Rapa Nui (Easter Island) in the South Pacific subtropical gyre. Science of the Total Environment. 2017;**586**:430-437. DOI: 10.1016/j.scitotenv.2017.01.175

[65] Yonkos L, Friedel E, Perez-Reyes A, Ghosal S, Arthur C. Microplastics in four estuarine rivers in the Chesapeake Bay, USA. Environmental Science & Technology. 2014;**48**(24):14195-14202. DOI: 10.1021/es5036317

[66] Sutton R, Mason S, Stanek S, Willis-Norton E, Wren I, Box C. Microplastic contamination in the San Francisco Bay, California, USA. Marine Pollution Bulletin. 2016;**109**(1):230-235. DOI: 10.1016/j.marpolbul.2016.05.077

[67] Alfonso M, Scordo F, Seitz C, Manstretta G, Ronda A, Arias A, et al. First evidence of microplastics in nine lakes across Patagonia (South America). Science of the Total Environment. 2020;**733**. DOI: 10.1016/j.scitotenv.2020.139385

[68] Jones K, Hartl M, Bell M, Capper A. Microplastic accumulation in a *Zostera marina* L. bed at Deerness sound, Orkney, Scotland. Marine Pollution Bulletin. 2020;**152**. DOI: 10.1016/j.marpolbul.2020.110883

[69] Kazour M, Jemaa S, Issa C, Khalaf G, Amara R. Microplastics pollution along the Lebanese coast (Eastern Mediterranean Basin): Occurrence in surface water, sediments and biota samples. Science of the Total Environment. 2019;**696**. DOI: 10.1016/j.scitotenv.2019.133933

[70] Güven O, Gökdag K, Jovanovic B, Kideys A. Microplastic litter composition of the Turkish territorial waters of the Mediterranean Sea, and its occurrence in the gastrointestinal tract of fish. Environmental Pollution. 2017;**223**:286-294. DOI: 10.1016/j.envpol.2017.01.025

[71] Nel H, Froneman P. A quantitative analysis of microplastic pollution along the south-eastern coastline of South Africa. Marine Pollution Bulletin. 2015;**101**(1):274-279. DOI: 10.1016/j.marpolbul.2015.09.043

[72] Cai M, He H, Liu M, Li S, Tang G, Wang W, et al. Lost but can't be neglected: Huge quantities of small microplastics hide in the South China Sea. Science of the Total Environment. 2018;**633**:1206-1216. DOI: 10.1016/j.scitotenv.2018.03.197

[73] Chen M, Jin M, Tao P, Wang Z, Xie W, Yu X, et al. Assessment of microplastics derived from mariculture in Xiangshan Bay, China. Environmental Pollution. 2018;**242**:1146-1156. DOI: 10.1016/j.envpol.2018.07.133

[74] Li J, Green C, Reynolds A, Shi H, Rotchell J. Microplastics in mussels sampled from coastal waters and supermarkets in the United Kingdom. Environmental Pollution. 2018;**241**:35-44. DOI: 10.1016/j.envpol.2018.05.038

[75] Sun J, Dai X, Wang Q, van Loosdrecht M, Ni B. Microplastics in wastewater treatment plants: Detection, occurrence and removal. Water Research. 2019;**152**:21-37. DOI: 10.1016/j.watres.2018.12.050

[76] Eo S, Hong S, Song Y, Han G, Seo S, Shim W. Prevalence of small high-density microplastics in the continental shelf and deep sea waters of East Asia.

Water Research. 2021;**200**. DOI: 10.1016/j.watres.2021.117238

[77] Kobayashi T, Yagi M, Kawaguchi T, Hata T, Shimizu K. Spatiotemporal variations of surface water microplastics near Kyushu, Japan: A quali-quantitative analysis. Marine Pollution Bulletin. 2021;**169**. DOI: 10.1016/j.marpolbul.2021.112563

[78] Samandra S, Johnston J, Jaeger J, Symons B, Xie S, Currell M, et al. Microplastic contamination of an unconfined groundwater aquifer in Victoria, Australia. Science of the Total Environment. 2022;**802**. DOI: 10.1016/j.scitotenv.2021.149727

[79] Leads R, Weinstein J. Occurrence of tire wear particles and other microplastics within the tributaries of the Charleston Harbor estuary, South Carolina, USA. Marine Pollution Bulletin. 2019;**145**:569-582. DOI: 10.1016/j.marpolbul.2019.06.061

[80] Martin J, Lusher A, Thompson R, Morley A. The deposition and accumulation of microplastics in marine sediments and bottom water from the Irish continental shelf. Scientific Reports. 2017;7. DOI: 10.1038/s41598-017-11079-2

[81] Scherer C, Weber A, Stock F, Vurusic S, Egerci H, Kochleus C, et al. Comparative assessment of microplastics in water and sediment of a large European river. Science of the Total Environment. 2020;**738**. DOI: 10.1016/j.scitotenv.2020.139866

[82] Leslie H, Brandsma S, van Velzen M, Vethaak A. Microplastics en route: Field measurements in the Dutch river delta and Amsterdam canals, wastewater treatment plants, North Sea sediments and biota. Environment International. 2017;**101**:133-142. DOI: 10.1016/j.envint.2017.01.018

[83] Sruthy S, Ramasamy E. Microplastic pollution in Vembanad Lake, Kerala, India: The first report of microplastics in lake and estuarine sediments in India. Environmental Pollution. 2017;**222**: 315-322. DOI: 10.1016/j.envpol.2016.12.038

[84] Ashwini S, Varghese G. Environmental forensic analysis of the microplastic pollution at "Nattika" Beach, Kerala coast, India. Environmental Forensics. 2020; **21**(1):21-36. DOI: 10.1080/15275922.2019.1693442

[85] Dodson G, Shotorban A, Hatcher P, Waggoner D, Ghosal S, Noffke N. Microplastic fragment and fiber contamination of beach sediments from selected sites in Virginia and North Carolina, USA. Marine Pollution Bulletin. 2020;**151**. DOI: 10.1016/j.marpolbul.2019.110869

[86] Lots F, Behrens P, Vijver M, Horton A, Bosker T. A large-scale investigation of microplastic contamination: Abundance and characteristics of microplastics in European beach sediment. Marine Pollution Bulletin. 2017;**123**(1–2): 219-226. DOI: 10.1016/j.marpolbul.2017.08.057

[87] Vaughan R, Turner S, Rose N. Microplastics in the sediments of a UK urban lake. Environmental Pollution. 2017;**229**:10-18. DOI: 10.1016/j.envpol.2017.05.057

[88] Rodrigues M, Abrantes N, Gonçalves F, Nogueira H, Marques J, Gonçalves A. Spatial and temporal distribution of microplastics in water and sediments of a freshwater system (Antua River, Portugal). Science of the

Total Environment. 2018;**633**:1549-1559. DOI: 10.1016/j.scitotenv.2018.03.233

[89] Abidli S, Toumi H, Lahbib Y, El Menif N. The first evaluation of microplastics in sediments from the complex Lagoon-Channel of Bizerte (Northern Tunisia). Water Air And Soil Pollution. 2017;**228**(7). DOI: 10.1007/s11270-017-3439-9

[90] Zobkov M, Esiukova E. Microplastics in Baltic bottom sediments: Quantification procedures and first results. Marine Pollution Bulletin. 2017;**114**(2):724-732. DOI: 10.1016/j.marpolbul.2016.10.060. Available from: https://pubmed.ncbi.nlm.nih.gov/27810093/

[91] Tanchuling MAN, Osorio ED. The microplastics in metro Manila Rivers: Characteristics, sources, and abatement. In: Stock F, Reifferscheid G, Brennholt N, Kostianaia E, editors. Plastics in the Aquatic Environment - Part I: Current Status and Challenges. Cham: Springer International Publishing; 2022. pp. 405-426

[92] Perez-Garcia P, Chow J, Costanzi E, Gurschke M, Dittrich J, Dierkes R, et al. An archaeal lid-containing feruloyl esterase degrades polyethylene terephthalate. Communications Chemistry. 2023;**6**(1). DOI: 10.1038/s42004-023-00998-z

[93] Lamichhane G, Acharya A, Marahatha R, Modi B, Paudel R, Adhikari A, et al. Microplastics in environment: Global concern, challenges, and controlling measures. International journal of Environmental Science and Technology. 2023;**20**(4):4673-4694. DOI: 10.1007/s13762-022-04261-1

[94] Rosal R. Morphological description of microplastic particles for environmental fate studies. Marine Pollution Bulletin. 2021;**171**. DOI: 10.1016/j.marpolbul.2021.112716

[95] Kooi M, Koelmans A. Simplifying microplastic via continuous probability distributions for size, shape, and density. Environmental Science & Technology Letters. 2019;**6**(9):551-557. DOI: 10.1021/acs.estlett.9b00379

[96] Chen D, Wang P, Liu S, Wang R, Wu Y, Zhu AX, et al. Global patterns of lake microplastic pollution: Insights from regional human development levels. Science of the Total Environment. 2024; **954**:176620. DOI: 10.1016/j.scitotenv.2024.176620

[97] Kabir MS, Wang H, Luster-Teasley S, Zhang LF, Zhao RZ. Microplastics in landfill leachate: Sources, detection, occurrence, and removal. Environmental Science and Ecotechnology. 2023;**16**:12. DOI: 10.1016/j.ese.2023.100256

[98] Simongini C, Pucetaite M, Serranti S, van Praagh M, Hammer E, Bonifazi G. Microplastics identification In landfill leachates by different spectroscopic techniques. Detritus. 2022;**18**:58-69. DOI: 10.31025/2611-4135/2022.15169

[99] Golwala H, Zhang XY, Iskander SM, Smith AL. Solid waste: An overlooked source of microplastics to the environment. Science of the Total Environment. 2021;**769**. DOI: 10.1016/j.scitotenv.2020.144581

[100] Yu F, Wu ZJ, Wang JY, Li YY, Chu RD, Pei YZ, et al. Effect of landfill age on the physical and chemical characteristics of waste plastics/microplastics in a waste landfill sites. Environmental Pollution. 2022;**306**. DOI: 10.1016/j.envpol.2022.119366

[101] Shen MC, Song B, Zhu Y, Zeng GM, Zhang YX, Yang YY, et al. Removal of microplastics via drinking water

treatment: Current knowledge and future directions. Chemosphere. 2020; **251**. DOI: 10.1016/j.chemosphere.2020.126612

[102] Shen MC, Xiong WP, Song B, Zhou CY, Almatrafi E, Zeng GM, et al. Microplastics in landfill and leachate: Occurrence, environmental behavior and removal strategies. Chemosphere. 2022;**305**. DOI: 10.1016/j.chemosphere.2022.135325

[103] Gossmann I, Herzke D, Held A, Schulz J, Nikiforov V, Georgi C, et al. Occurrence and backtracking of microplastic mass loads including tire wear particles in northern Atlantic air. Nature Communications. 2023; **14**(1):9. DOI: 10.1038/s41467-023-39340-5

[104] Karbalaei S, Hanachi P, Walker TR, Cole M. Occurrence, sources, human health impacts and mitigation of microplastic pollution. Environmental Science and Pollution Research. 2018; **25**(36):36046-36063. DOI: 10.1007/s11356-018-3508-7

[105] Torres-Agullo A, Karanasiou A, Moreno T, Lacorte S. Overview on the occurrence of microplastics in air and implications from the use of face masks during the COVID-19 pandemic. Science of the Total Environment. 2021; **800**. DOI: 10.1016/j.scitotenv.2021.149555

[106] Welden NA, Lusher A. Chapter 9 - Microplastics: From origin to impacts. In: Letcher TM, editor. Plastic Waste and Recycling. Academic Press; 2020

[107] Keshta AE, Shaltout KH, Baldwin AH, El-Din AS, Eid EM. Variation in plant community composition and biomass to macro and micronutrients and salinity across Egypt's five major Coastal Lakes. Sustainability. 2022;**14**(10):15. DOI: 10.3390/su14106180

[108] Nava V, Chandra S, Aherne J, Alfonso MB, Antao-Geraldes AM, Attermeyer K, et al. Plastic debris in lakes and reservoirs. Nature. 2023; **619**(7969):317. DOI: 10.1038/s41586-023-06168-4

[109] Huang YL, Tian M, Jin F, Chen MY, Liu ZG, He SQ, et al. Coupled effects of urbanization level and dam on microplastics in surface waters in a coastal watershed of Southeast China. Marine Pollution Bulletin. 2020;**154**. DOI: 10.1016/j.marpolbul.2020.111089

[110] Kukkola AT, Senior G, Silburn B, Bakir A, Maes T, Kroger S, et al. A large-scale study of microplastic abundance in sediment cores from the UK continental shelf and slope. Marine Pollution Bulletin. 2022;**178**. DOI: 10.1016/j.marpolbul.2022.113554

[111] Turner S, Horton AA, Rose NL, Hall C. A temporal sediment record of microplastics in an urban lake, London, UK. Journal of Paleolimnology. 2019; **61**(4):449-462. DOI: 10.1007/s10933-019-00071-7

[112] D'Avignon G, Gregory-Eaves I, Ricciardi A. Microplastics in lakes and rivers: An issue of emerging significance to limnology. Environmental Reviews. 2022;**30**(2):228-244. DOI: 10.1139/er-2021-0048

[113] Su L, Xue YG, Li LY, Yang DQ, Kolandhasamy P, Li DJ, et al. Microplastics in Taihu Lake, China. Environmental Pollution. 2016; **216**:711-719. DOI: 10.1016/j.envpol.2016.06.036

[114] Horton AA, Svendsen C, Williams RJ, Spurgeon DJ, Lahive E.

Large microplastic particles in sediments of tributaries of the river Thames, UK - abundance, sources and methods for effective quantification. Marine Pollution Bulletin. 2017;**114**(1):218-226. DOI: 10.1016/j.marpolbul.2016.09.004

[115] Zurier HS, Goddard JM. Biodegradation of microplastics in food and agriculture. Current Opinion in Food Science. 2021;**37**:37-44. DOI: 10.1016/j.cofs.2020.09.001

[116] Talbot R, Chang HJ. Microplastics in freshwater: A global review of factors affecting spatial and temporal variations*. Environmental Pollution. 2022;**292**. DOI: 10.1016/j.envpol.2021.118393

[117] Kwon JH, Kim JW, Pham TD, Tarafdar A, Hong S, Chun SH, et al. Microplastics in food: A review on analytical methods and challenges. International Journal of Environmental Research and Public Health. 2020;**17**(18). DOI: 10.3390/ijerph17186710

[118] Pal D, Prabhakar R, Barua VB, Zekker I, Burlakovs J, Krauklis A, et al. Microplastics in aquatic systems: A comprehensive review of its distribution, environmental interactions, and health risks. Environmental Science and Pollution Research. 2024;**32**:56-88. DOI: 10.1007/s11356-024-35741-1

[119] Greenshields J, Anastasi A, Irving AD, Capper A. A systematic review to assess current surface water and sediment microplastic sampling practices in seagrass and mangrove ecosystems. Environmental Science and Pollution Research. 2024;**31**:66615-66629. DOI: 10.1007/s11356-024-35690-9

[120] Summers E, Du J, Park K, Wharton M, Kaiser K. Importance of the water-sediment bed interactions in simulating microplastic particles in an estuarine system. Frontiers in Marine Science. 2024;**11**. DOI: 10.3389/fmars.2024.1414459

[121] Stock F, Kochleus C, Bansch-Baltruschat B, Brennholt N, Reifferscheid G. Sampling techniques and preparation methods for microplastic analyses in the aquatic environment - A review. Trac-Trends in Analytical Chemistry. 2019;**113**:84-92. DOI: 10.1016/j.trac.2019.01.014

[122] McColley CJ, Nason JA, Harper BJ, Harper SL. An assessment of methods used for the generation and characterization of cryomilled polystyrene micro- and nanoplastic particles. Microplastics and Nanoplastics. 2023;**3**(1):20. DOI: 10.1186/s43591-023-00069-z

[123] Yee MSL, Hii LW, Looi CK, Lim WM, Wong SF, Kok YY, et al. Impact of microplastics and Nanoplastics on human health. Nanomaterials. 2021;**11**(2). DOI: 10.3390/nano11020496

[124] Campanale C, Massarelli C, Savino I, Locaputo V, Uricchio VF. A detailed review study on potential effects of microplastics and additives of concern on human health. International Journal of Environmental Research and Public Health. 2020;**17**(4). DOI: 10.3390/ijerph17041212

[125] Yuan ZH, Nag R, Cummins E. Human health concerns regarding microplastics in the aquatic environment- From marine to food systems. Science of the Total Environment. 2022;**823**. DOI: 10.1016/j.scitotenv.2022.153730

[126] López APA, Trilleras J, Arana VA, Garcia-Alzate LS, Grande-Tovar CD. Atmospheric microplastics: Exposure, toxicity, and detrimental health effects.

RSC Advances. 2023;**13**(11):7468-7489. DOI: 10.1039/d2ra07098g

[127] Kannan K, Vimalkumar K. A review of human exposure to microplastics and insights into microplastics as obesogens. Frontiers in Endocrinology. 2021;**12**. DOI: 10.3389/fendo.2021.724989

[128] Huang Z, Hu B, Wang H. Analytical methods for microplastics in the environment: A review. Environmental Chemistry Letters. 2023;**21**(1):383-401. DOI: 10.1007/s10311-022-01525-7

[129] Kumar R, Verma A, Shome A, Sinha R, Sinha S, Jha PK, et al. Impacts of plastic pollution on ecosystem services, sustainable development goals, and need to focus on circular economy and policy interventions. Sustainability. 2021;**13** (17). DOI: 10.3390/su13179963

[130] Akdogan Z, Guven B. Microplastics in the environment: A critical review of current understanding and identification of future research needs. Environmental Pollution. 2019;**254**. DOI: 10.1016/j.envpol.2019.113011

[131] van den Berg P, Huerta-Lwanga E, Corradini F, Geissen V. Sewage sludge application as a vehicle for microplastics in eastern Spanish agricultural soils. Environmental Pollution. 2020;**261**:7. DOI: 10.1016/j.envpol.2020.114198

[132] Boots B, Russell CW, Green DS. Effects of microplastics in soil ecosystems: Above and below ground. Environmental Science & Technology. 2019;**53**(19):11496-11506. DOI: 10.1021/acs.est.9b03304

[133] Lwanga EH, Gertsen H, Gooren H, Peters P, Salánki T, van der Ploeg M, et al. Microplastics in the terrestrial ecosystem: Implications for *Lumbricus terrestris* (Oligochaeta, Lumbricidae). Environmental Science & Technology. 2016;**50**(5):2685-2691. DOI: 10.1021/acs.est.5b05478

[134] Horton A, Walton A, Spurgeon D, Lahive E, Svendsen C. Microplastics in freshwater and terrestrial environments: Evaluating the current understanding to identify the knowledge gaps and future research priorities. Science of the Total Environment. 2017;**586**:127-141. DOI: 10.1016/j.scitotenv.2017.01.190

[135] Tanaka K, Watanuki Y, Takada H, Ishizuka M, Yamashita R, Kazama M, et al. *In vivo* accumulation of plastic-derived chemicals into seabird tissues. Current Biology. 2020;**30**(4):723. DOI: 10.1016/j.cub.2019.12.037

[136] Hou D, Rao D. Microplastics: Their effects on amphibians and reptiles-A review. Pakistan Journal of Zoology. 2022;**54**(6):2931-2951. DOI: 10.17582/journal.pjz/20210820080823

[137] Najibzadeh M, Kazemi A, Ul Hassan H, Esmaeilbeigi M. Hazard assessment of microplastics and heavy metals contamination in Levant frogs (*Pelophylax bedriagae*): A bioindicator in Western Iran. Environmental Research. 2024;**262**. DOI: 10.1016/j.envres.2024.119774

[138] Machado AAD, Kloas W, Zarfl C, Hempel S, Rillig MC. Microplastics as an emerging threat to terrestrial ecosystems. Global Change Biology. 2018;**24**(4):1405-1416. DOI: 10.1111/gcb.14020

[139] McCormick A, Hoellein TJ, Mason SA, Schluep J, Kelly JJ. Microplastic is an abundant and distinct microbial habitat in an Urban River. Environmental Science & Technology. 2014;**48**(20): 11863-11871. DOI: 10.1021/es503610r

[140] Nizzetto L, Futter M, Langaas S. Are agricultural soils dumps for

microplastics of urban origin? Environmental Science & Technology. 2016;**50**(20):10777-10779. DOI: 10.1021/acs.est.6b04140

[141] Li JY, Liu HH, Chen JP. Microplastics in freshwater systems: A review on occurrence, environmental effects, and methods for microplastics detection. Water Research. 2018;**137**: 362-374. DOI: 10.1016/j.watres.2017.12.056

[142] Zhang XL, Liu L, Chen XL, Li J, Chen JX, Wang HJ. The fate and risk of microplastic and antibiotic sulfamethoxazole coexisting in the environment. Environmental Geochemistry and Health. 2023;**45**(6): 2905-2915. DOI: 10.1007/s10653-022-01385-8

[143] Cui WZ, Gao PP, Zhang MY, Wang L, Sun HW, Liu CG. Adverse effects of microplastics on earthworms: A critical review. Science of the Total Environment. 2022;**850**. DOI: 10.1016/j.scitotenv.2022.158041

[144] Kim SW, An YJ. Edible size of polyethylene microplastics and their effects on springtail behavior. Environmental Pollution. 2020;**266**:6. DOI: 10.1016/j.envpol.2020.115255

[145] Ji ZY, Huang Y, Feng Y, Johansen A, Xue JM, Tremblay LA, et al. Effects of pristine microplastics and nanoplastics on soil invertebrates: A systematic review and meta-analysis of available data. Science of the Total Environment. 2021;**788**:13. DOI: 10.1016/j.scitotenv.2021.147784

[146] Kershaw PJ, Rochman CM. Sources, fate and effects of microplastics in the marine environment: Part 2 of a global assessment. Reports and studies-IMO/FAO/Unesco-IOC/WMO/IAEA/UN/UNEP joint group of experts on the scientific aspects of marine environmental protection (GESAMP) Eng No. 93. 2015

[147] Ng EL, Lin SY, Dungan AM, Colwell JM, Ede S, Lwanga EH, et al. Microplastic pollution alters forest soil microbiome. Journal of Hazardous Materials. 2021;**409**. DOI: 10.1016/j.jhazmat.2020.124606

[148] Zhao TT, Lozano YM, Rillig MC. Microplastics increase soil pH and decrease microbial activities as a function of microplastic shape, polymer type, and exposure time. Frontiers in Environmental Science. 2021;**9**:14. DOI: 10.3389/fenvs.2021.675803

[149] de Sá LC, Oliveira M, Ribeiro F, Rocha TL, Futter MN. Studies of the effects of microplastics on aquatic organisms: What do we know and where should we focus our efforts in the future? Science of the Total Environment. 2018; **645**:1029-1039. DOI: 10.1016/j.scitotenv.2018.07.207

[150] Sayed A, Hamed M, Badrey AEA, Ismail RF, Osman YAA, Osman AGM, et al. Microplastic distribution, abundance, and composition in the sediments, water, and fishes of the red and Mediterranean seas, Egypt. Marine Pollution Bulletin. 2021;**173**. DOI: 10.1016/j.marpolbul.2021.112966

[151] Catarino AI, Macchia V, Sanderson WG, Thompson RC, Henry TB. Low levels of microplastics (MP) in wild mussels indicate that MP ingestion by humans is minimal compared to exposure via household fibres fallout during a meal. Environmental Pollution. 2018;**237**: 675-684. DOI: 10.1016/j.envpol.2018.02.069

[152] Barboza LGA, Dick Vethaak A, Lavorante BRBO, Lundebye A-K,

Guilhermino L. Marine microplastic debris: An emerging issue for food security, food safety and human health. Marine Pollution Bulletin. 2018; **133**:336-348. DOI: 10.1016/j.marpolbul. 2018.05.047

[153] Jemec A, Horvat P, Kunej U, Bele M, Krzan A. Uptake and effects of microplastic textile fibers on freshwater crustacean *Daphnia magna*. Environmental Pollution. 2016;**219**: 201-209. DOI: 10.1016/j. envpol.2016.10.037

[154] Beyer J, Green N, Brooks S, Allan I, Ruus A, Gomes T, et al. Blue mussels (*Mytilus edulis* spp.) as sentinel organisms in coastal pollution monitoring: A review. Marine Environmental Research. 2017;**130**: 338-365. DOI: 10.1016/j. marenvres.2017.07.024

[155] Lee JH, Kang JC, Kim JH. Toxic effects of microplastic (polyethylene) on fish: Accumulation, hematological parameters and antioxidant responses in Korean bullhead, *Pseudobagrus fulvidraco*. Science of the Total Environment. 2023;**877**. DOI: 10.1016/j. scitotenv.2023.162874

[156] Das S, Chandrasekaran N, Mukherjee A. Unmasking effects of masks: Microplastics released from disposable surgical face masks induce toxic effects in microalgae *Scenedesmus obliquus* and chlorella sp. Comparative Biochemistry and Physiology C-Toxicology & Pharmacology. 2023;**267**. DOI: 10.1016/j. cbpc.2023.109587

[157] Pacheco A, Martins A, Guilhermino L. Toxicological interactions induced by chronic exposure to gold nanoparticles and microplastics mixtures in *Daphnia magna*. Science of the Total Environment. 2018;**628-629**:474-483. DOI: 10.1016/j.scitotenv.2018.02.081

[158] He MT, Yan MT, Chen XF, Wang XK, Gong H, Wang WJ, et al. Bioavailability and toxicity of microplastics to zooplankton. Gondwana Research. 2022;**108**:120-126. DOI: 10.1016/j.gr.2021.07.021

[159] Jovanovic B. Ingestion of microplastics by fish and its potential consequences from a physical perspective. Integrated Environmental Assessment and Management. 2017; **13**(3):510-515. DOI: 10.1002/ieam.1913

[160] Sussarellu R, Suquet M, Thomas Y, Lambert C, Fabioux C, Pernet MEJ, et al. Oyster reproduction is affected by exposure to polystyrene microplastics. Proceedings of the National Academy of Sciences of the United States of America. 2016;**113**(9):2430-2435. DOI: 10.1073/ pnas.1519019113

[161] Singleman C, Zimmerman A, Harrison E, Roy NK, Wirgin I, Holtzman NG. Toxic effects of polychlorinated biphenyl congeners and Aroclors on embryonic growth and development. Environmental Toxicology and Chemistry. 2021;**40**(1):187-201. DOI: 10.1002/etc.4908

[162] Rivers-Auty J, Bond AL, Grant ML, Lavers JL. The one-two punch of plastic exposure: Macro- and micro-plastics induce multi-organ damage in seabirds. Journal of Hazardous Materials. 2023; **442**. DOI: 10.1016/j.jhazmat.2022. 130117

[163] Zhang XN, Zhang H, Yu K, Li N, Liu YN, Liu XY, et al. Rapid monitoring approach for microplastics using portable pyrolysis-mass spectrometry. Analytical Chemistry. 2020;**92**(6): 4656-4662. DOI: 10.1021/acs.analchem. 0c00300

[164] Vo HC, Pham MH. Ecotoxicological effects of microplastics on aquatic organisms: A review. Environmental Science and Pollution Research. 2021; **28**(33):44716-44725. DOI: 10.1007/s11356-021-14982-4

[165] Jiang XF, Chang YQ, Zhang T, Qiao Y, Klobucar G, Li M. Toxicological effects of polystyrene microplastics on earthworm (*Eisenia fetida*). Environmental Pollution. 2020;**259**. DOI: 10.1016/j.envpol.2019.113896

[166] Allen S, Allen D, Phoenix VR, Le Roux G, Jiménez PD, Simonneau A, et al. Atmospheric transport and deposition of microplastics in a remote mountain catchment. Nature Geoscience. 2019; **12**(5):339. DOI: 10.1038/s41561-019-0335-5

[167] Gasperi J, Wright SL, Dris R, Collard F, Mandin C, Guerrouache M, et al. Microplastics in air: Are we breathing it in? Current Opinion in Environmental Science & Health. 2018;**1**: 1-5. DOI: 10.1016/j.coesh.2017.10.002

[168] Wright SL, Kelly FJ. Plastic and human health: A micro issue? Environmental Science & Technology. 2017;**51**(12):6634-6647. DOI: 10.1021/acs.est.7b00423

[169] Mahler GJ, Esch MB, Tako E, Southard TL, Archer SD, Glahn RP, et al. Oral exposure to polystyrene nanoparticles affects iron absorption. Nature Nanotechnology. 2012;**7**(4):264-U1500. DOI: 10.1038/nnano.2012.3

[170] Prata JC, da Costa JP, Lopes I, Andrady AL, Duarte AC, Rocha-Santos T. A one health perspective of the impacts of microplastics on animal, human and environmental health. Science of the Total Environment. 2021; **777**. DOI: 10.1016/j.scitotenv.2021.146094

[171] Merrill WW. Flock worker's lung. Annals of Internal Medicine. 1999; **130**(7):615-615. DOI: 10.7326/0003-4819-130-7-199904060-00010

[172] Umasankar U, Sabumon PC. Microplastics in plant species: Impacts and ecological perspectives. In: Egbueri JC, Ighalo JO, Pande CB, editors. Microplastics in African and Asian Environments: The Influencers, Challenges, and Solutions. Springer Nature Switzerland: Cham; 2024. pp. 331-359

[173] Wang Z, Liu S, Cheng Z, Xu G, Li F, Bu Q, et al. Endoplasmic reticulum stress exacerbates microplastics-induced toxicity in animal cells. Food Research International. 2024;**175**. DOI: 10.1016/j.foodres.2023.113818

[174] Parker L. We made plastic. We depend on it. Now we're drowning in it. Available from: https://www.nationalgeographic.com/magazine/article/plastic-planet-waste-pollution-trash-crisis

[175] Hopewell J, Dvorak R, Kosior E. Plastics recycling: Challenges and opportunities. Philosophical Transactions of the Royal Society B-Biological Sciences. 2009;**364**(1526): 2115-2126. DOI: 10.1098/rstb.2008.0311

[176] Lambert S, Wagner M. Environmental performance of bio-based and biodegradable plastics: The road ahead. Chemical Society Reviews. 2017;**46**(22):6855-6871. DOI: 10.1039/c7cs00149e

[177] Lau WWY, Shiran Y, Bailey RM, Cook E, Stuchtey MR, Koskella J, et al. Evaluating scenarios toward zero plastic pollution. Science. 2020;**369**(6510):1455. DOI: 10.1126/science.aba9475

[178] Rios LM, Moore C, Jones PR. Persistent organic pollutants carried by

synthetic polymers in the ocean environment. Marine Pollution Bulletin. 2007;**54**(8):1230-1237. DOI: 10.1016/j.marpolbul.2007.03.022

[179] Leiser R, Wu GM, Neu TR, Wendt-Potthoff K. Biofouling, metal sorption and aggregation are related to sinking of microplastics in a stratified reservoir. Water Research. 2020;**176**. DOI: 10.1016/j.watres.2020.115748

[180] Sundbaek KB, Koch IDW, Villaro CG, Rasmussen NS, Holdt SL, Hartmann NB. Sorption of fluorescent polystyrene microplastic particles to edible seaweed Fucus vesiculosus. Journal of Applied Phycology. 2018; **30**(5):2923-2927. DOI: 10.1007/s10811-018-1472-8

[181] Zheng ZJ, Huang Y, Liu LA, Wang L, Tang JC. Interaction between microplastic biofilm formation and antibiotics: Effect of microplastic biofilm and its driving mechanisms on antibiotic resistance gene. Journal of Hazardous Materials. 2023;**459**. DOI: 10.1016/j.jhazmat.2023.132099

[182] Martins MJF, Mota CF, Pearson GA. Sex-biased gene expression in the brown alga Fucus vesiculosus. BMC Genomics. 2013;**14**. DOI: 10.1186/1471-2164-14-294

[183] Nolte TM, Hartmann NB, Kleijn JM, Garnaes J, van de Meent D, Hendriks AJ, et al. The toxicity of plastic nanoparticles to green algae as influenced by surface modification, medium hardness and cellular adsorption. Aquatic Toxicology. 2017;**183**:11-20. DOI: 10.1016/j.aquatox.2016.12.005

[184] Acarer S. A review of microplastic removal from water and wastewater by membrane technologies. Water Science and Technology. 2023;**88**(1):199-219. DOI: 10.2166/wst.2023.186

[185] Horton AA, Dixon SJ. Microplastics: An introduction to environmental transport processes. Wiley Interdisciplinary Reviews-Water. 2018;**5**(2). DOI: 10.1002/wat2.1268

[186] Yi KX, Huang JH, Li X, Li SZ, Pang HL, Liu ZX, et al. Long-term impacts of polyethylene terephthalate (PET) microplastics in membrane bioreactor. Journal of Environmental Management. 2022;**323**. DOI: 10.1016/j.jenvman.2022.116234

[187] Padervand M, Lichtfouse E, Robert D, Wang CY. Removal of microplastics from the environment. A review. Environmental Chemistry Letters. 2020;**18**(3):807-828. DOI: 10.1007/s10311-020-00983-1

[188] Harrison JP, Sapp M, Schratzberger M, Osborn AM. Interactions between microorganisms and marine microplastics: A call for research. Marine Technology Society Journal. 2011;**45**(2):12-20. DOI: 10.4031/mtsj.45.2.2

[189] Khandare SD, Chaudhary DR, Jha B. Marine bacterial biodegradation of low-density polyethylene (LDPE) plastic. Biodegradation. 2021;**32**(2): 127-143. DOI: 10.1007/s10532-021-09927-0

[190] Dawson AL, Kawaguchi S, King CK, Townsend KA, King R, Huston WM, et al. Turning microplastics into nanoplastics through digestive fragmentation by Antarctic krill. Nature Communications. 2018;**9**. DOI: 10.1038/s41467-018-03465-9

[191] Solanki S, Sinha S, Singh R. Myco-degradation of microplastics: An account of identified pathways and analytical methods for their determination. Biodegradation. 2022;**33**(6):529-556. DOI: 10.1007/s10532-022-10001-6

[192] So MWK, Vorsatz LD, Cannicci S, Not C. The role of mangrove crabs, the key macrofaunal bioengineers, in microplastic production in tropical coastal forests. Regional Studies in Marine Science. 2023;**63**. DOI: 10.1016/j.rsma.2023.103012

[193] Maity W, Maity S, Bera S, Roy A. Emerging roles of PETase and MHETase in the biodegradation of plastic wastes. Applied Biochemistry and Biotechnology. 2021;**193**(8):2699-2716. DOI: 10.1007/s12010-021-03562-4

[194] Cai ZM, Li MQ, Zhu ZY, Wang XC, Huang YY, Li TM, et al. Biological degradation of plastics and microplastics: A recent perspective on associated mechanisms and influencing factors. Microorganisms. 2023;**11**(7). DOI: 10.3390/microorganisms11071661

[195] Al Hosni AS, Pittman JK, Robson GD. Microbial degradation of four biodegradable polymers in soil and compost demonstrating polycaprolactone as an ideal compostable plastic. Waste Management. 2019;**97**:105-114. DOI: 10.1016/j.wasman.2019.07.042

[196] Ngo PL, Pramanik BK, Shah K, Roychand R. Pathway, classification and removal efficiency of microplastics in wastewater treatment plants. Environmental Pollution. 2019;**255**. DOI: 10.1016/j.envpol.2019.113326

[197] Shen MC, Zhang YX, Almatrafi E, Hu T, Zhou CY, Song B, et al. Efficient removal of microplastics from wastewater by an electrocoagulation process. Chemical Engineering Journal. 2022;**428**. DOI: 10.1016/j.cej.2021.131161

[198] Rajala K, Gronfors O, Hesampour M, Mikola A. Removal of microplastics from secondary wastewater treatment plant effluent by coagulation/flocculation with iron, aluminum and polyamine-based chemicals. Water Research. 2020;**183**. DOI: 10.1016/j.watres.2020.116045

[199] Lapointe M, Farner JM, Hernandez LM, Tufenkji N. Understanding and improving microplastic removal during water treatment: Impact of coagulation and flocculation. Environmental Science & Technology. 2020;**54**(14):8719-8727. DOI: 10.1021/acs.est.0c00712

[200] Sturm MT, Horn H, Schuhen K. Removal of microplastics from waters through agglomeration-fixation using Organosilanes-effects of polymer types, water composition and temperature. Water. 2021;**13**(5). DOI: 10.3390/w13050675

[201] Fadli MH, Ibadurrohman M, Slamet S. Microplastic pollutant degradation in water using modified TiO_2 photocatalyst under UV-irradiation. In: IOP Conference Series: Materials Science and Engineering, 2021. Bristol, United Kingdom: IOP Publishing; 2021. p. 012055

[202] Nabi I, Bacha AUR, Li KJ, Cheng HY, Wang T, Liu YY, et al. Complete photocatalytic mineralization of microplastic on TiO_2 nanoparticle film. Iscience. 2020;**23**(7). DOI: 10.1016/j.isci.2020.101326

[203] Liu K, Wang XH, Fang T, Xu P, Zhu LX, Li DJ. Source and potential risk assessment of suspended atmospheric microplastics in Shanghai. Science of the Total Environment. 2019;**675**:462-471. DOI: 10.1016/j.scitotenv.2019.04.110

[204] Liu F, Nord NB, Bester K, Vollertsen J. Microplastics removal from

treated wastewater by a biofilter. Water. 2020;**12**(4). DOI: 10.3390/w12041085

[205] Arossa S, Martin C, Rossbach S, Duarte CM. Microplastic removal by Red Sea giant clam (*Tridacna maxima*). Environmental Pollution. 2019;**252**: 1257-1266. DOI: 10.1016/j.envpol.2019.05.149

[206] Chen Z, Zhao WQ, Xing RZ, Xie SJ, Yang XG, Cui P, et al. Enhanced in situ biodegradation of microplastics in sewage sludge using hyperthermophilic composting technology. Journal of Hazardous Materials. 2020;**384**. DOI: 10.1016/j.jhazmat.2019.121271

[207] Osman HEM, Elaidarous AA, El-Morsy MH, Eid EM, Keshta AE. Soils with more clay and dense vegetation were rich in soil carbon along Wadi Al-Sharaea, Makkah, Saudi Arabia. Heliyon. 2023;**9**(1):e12988. DOI: 10.1016/j.heliyon.2023.e12988

[208] Essa DI, Elshobary ME, Attiah AM, Salem ZE, Keshta AE, Edokpayi JN. Assessing phytoplankton populations and their relation to water parameters as early alerts and biological indicators of the aquatic pollution. Ecological Indicators. 2024;**159**:111721. DOI: 10.1016/j.ecolind.2024.111721

[209] Shaheen ME, Gagnon JE, Barrette JC, Keshta AE. Evaluation of pollution levels in sediments from Lake Edku, Egypt using laser ablation inductively coupled plasma mass spectrometry. Marine Pollution Bulletin. 2024;**202**:116387. DOI: 10.1016/j.marpolbul.2024.116387

[210] Dajam AS, Keshta AE, Bindajam AA, Al-Qthanin RN, Arshad M, Eid EM. Mangrove (*Avicennia marina*) conservation contributed to a higher carbon sequestration rate at protected sites compared to overgrazed mangrove forests. Journal of Soil Science and Plant Nutrition. 2024;**24**:4868-4879. DOI: 10.1007/s42729-024-01878-8

[211] El-Hamid H, Kacem H, Keshta A, El-Alfy M. The economic benefit of coastal blue carbon stocks in Edku Lake: A case study at the Mediterranean Sea. South African Journal of Botany. 2024; **173**:76-84. DOI: 10.1016/j.sajb.2024.08.017

[212] Dajam A, Keshta A, Bindajam A, Eid E. Bioaccumulation of heavy metals in mangrove (*Avicennia marina*): Predictive uptake modeling and phytoremediation potential. Journal of Soil Science and Plant Nutrition. 2024; **24**:6085-6098. DOI: 10.1007/s42729-024-01962-z

[213] Bydalek F, Ifayemi D, Reynolds L, Barden R, Kasprzyk-Hordern B, Wenk J. Microplastic dynamics in a free water surface constructed wetland. Science of the Total Environment. 2023;**858**. DOI: 10.1016/j.scitotenv.2022.160113

[214] Cunningham EM, Seijo NR, Altieri KE, Audh RR, Burger JM, Bornman TG, et al. The transport and fate of microplastic fibres in the Antarctic: The role of multiple global processes. Frontiers in Marine Science. 2022;**9**. DOI: 10.3389/fmars.2022.1056081

[215] Quik JTK, Meesters JAJ, Koelmans AA. A multimedia model to estimate the environmental fate of microplastic particles. Science of the Total Environment. 2023;**882**. DOI: 10.1016/j.scitotenv.2023.163437

[216] Helcoski R, Yonkos LT, Sanchez A, Baldwin AH. Wetland soil microplastics are negatively related to vegetation cover and stem density. Environmental Pollution. 2020;**256**. DOI: 10.1016/j.envpol.2019.113391

[217] He BB, Wijesiri B, Ayoko GA, Egodawatta P, Rintoul L, Goonetilleke A. Influential factors on microplastics occurrence in river sediments. Science of the Total Environment. 2020;**738**. DOI: 10.1016/j.scitotenv.2020.139901

[218] Kadlec RH, Roy SB, Munson RK, Charlton S, Brownlie W. Water quality performance of treatment wetlands in the Imperial Valley, California. Ecological Engineering. 2010;**36**(8): 1093-1107. DOI: 10.1016/j.ecoleng. 2010.04.028

[219] Vymazal J. Constructed wetlands for wastewater treatment. Water. 2010; **2**(3):530-549. DOI: 10.3390/w2030530

[220] Chen YL, Li TC, Hu HJ, Ao HY, Xiong X, Shi HH, et al. Transport and fate of microplastics in constructed wetlands: A microcosm study. Journal of Hazardous Materials. 2021;**415**. DOI: 10.1016/j. jhazmat.2021.125615

[221] Burlakova LE, Karatayev AY, Padilla DK, Cartwright LD, Hollas DN. Wetland restoration and invasive species: Apple snail (*Pomacea insularum*) feeding on native and invasive aquatic plants. Restoration Ecology. 2009; **17**(3):433-440. DOI: 10.1111/ j.1526-100X.2008.00429.x

[222] Shaltout KH, El-Beheiry MA, El-Kady HF, Keshta AE. Distribution of the naturalized species *Dalbergia sissoo* Roxb. ex DC. in Nile Delta, Egypt. Taeckholmia. 2011;**30**(1):59-79. DOI: 10.21608/TAEC.2010.155345

[223] Shaltout KH, Keshta AE. The biology of Egyptian woody perennials 4. Dalbergia sissoo roxb. ex DC. Assiut University Bulletin for Environmental Researches. 2011;**14**(2):131-152

[224] Keshta AE, Yarwood SA, Baldwin AH. Methane emissions are hghly variable across wetland habitats in natural and restored tidal freshwater wetlands. Wetlands. 2023;**43**(5). DOI: 10.1007/s13157-023-01701-7

[225] Li RL, Zhang SP, Zhang LL, Yu KF, Wang SP, Wang YH. Field study of the microplastic pollution in sea snails (*Ellobium chinense*) from mangrove forest and their relationships with microplastics in water/sediment located on the north of Beibu Gulf. Environmental Pollution. 2020;**263**. DOI: 10.1016/j.envpol.2020.114368

[226] John J, Nandhini AR, Chellam PV, Sillanpaa M. Microplastics in mangroves and coral reef ecosystems: A review. Environmental Chemistry Letters. 2022; **20**(1):397-416. DOI: 10.1007/s10311-021-01326-4

[227] Ferrero L, Scibetta L, Markuszewski P, Mazurkiewicz M, Drozdowska V, Makuch P, et al. Airborne and marine microplastics from an oceanographic survey at the Baltic Sea: An emerging role of air-sea interaction? Science of the Total Environment. 2022;**824**. DOI: 10.1016/j. scitotenv.2022.153709

[228] Mohammadi A, Malakootian M, Dobaradaran S, Hashemi M, Jaafarzadeh N. Occurrence, seasonal distribution, and ecological risk assessment of microplastics and phthalate esters in leachates of a landfill site located near the marine environment: Bushehr port, Iran as a case. Science of the Total Environment. 2022;**842**. DOI: 10.1016/j.scitotenv. 2022.156838

[229] Hamed M, Osman AGM, Badrey AEA, Soliman HAM, Sayed AEH. Microplastics-induced Eryptosis and Poikilocytosis in early-juvenile Nile tilapia (*Oreochromis niloticus*). Frontiers

in Physiology. 2021;**12**. DOI: 10.3389/fphys.2021.742922

[230] Lambert S, Sinclair C, Boxall A. Occurrence, degradation, and effect of polymer-based materials in the environment. Reviews of Environmental Contamination and Toxicology. 2014; **227**(227):1-53. DOI: 10.1007/978-3-319-01327-5_1

[231] Deng LZ, Cai L, Sun FY, Li G, Che Y. Public attitudes towards microplastics: Perceptions, behaviors and policy implications. Resources Conservation and Recycling. 2020;**163**. DOI: 10.1016/j.resconrec.2020.105096

[232] Hermsen E, Mintenig SM, Besseling E, Koelmans AA. Quality criteria for the analysis of microplastic in biota samples: A critical review. Environmental Science & Technology. 2018;**52**(18):10230-10240. DOI: 10.1021/acs.est.8b01611

[233] Yu P, Liu ZQ, Wu DL, Chen MH, Lv WW, Zhao YL. Accumulation of polystyrene microplastics in juvenile *Eriocheir sinensis* and oxidative stress effects in the liver. Aquatic Toxicology. 2018;**200**:28-36. DOI: 10.1016/j.aquatox.2018.04.015

[234] Gao RR, Sun CM. A marine bacterial community capable of degrading poly(ethylene terephthalate) and polyethylene. Journal of Hazardous Materials. 2021;**416**. DOI: 10.1016/j.jhazmat.2021.125928

[235] Hong SH, Shim WJ, Hong L. Methods of analysing chemicals associated with microplastics: A review. Analytical Methods. 2017;**9**(9):1361-1368. DOI: 10.1039/c6ay02971j

Chapter 4

Physical Pre-Treatment of Renewable Fuels for Custom-Made Use in Fixed and Fluidized Bed Reactors

Maxwell Appiah and Prince L. Tetteh

Abstract

The physical pre-treatment of renewable fuels specifically woody biomass is critical in enhancing their efficiency and suitability for use in fixed and fluidized bed reactors. This process involves a series of modifications that improve fuel characteristics such as size, shape, moisture content, and homogeneity to optimize performance in energy conversion systems. In fixed bed reactors, where fuel particles remain stationary, ensuring uniform particle size and reducing moisture content can significantly enhance combustion efficiency and emission profiles. For fluidized bed reactors, where particles are suspended in a fluid-like state through the injection of air or gas, achieving an appropriate particle size distribution ensures superior heat transfer, reduced pollutant formation, and consistent reactor performance. This chapter will explore the varied techniques employed in the pre-treatment stages, with a focus on the milling techniques of size reduction, to tailor renewable fuels like biomass wastes, and residues for bespoke reactor applications. Knife mills are generally best suited for particle size reduction; they are about 25% more energy-efficient than hammer mills, and can produce 0.25 mm sizes of milled woody biomass which is suitable for biofuel production in reactors as compared to hammer and ball mills. The scope includes identifying specific challenges and opportunities associated with each size reduction method, ultimately providing a foundation for innovative solutions that accommodate diverse fuel types and reactor designs.

Keywords: woody biomass waste, size reduction, milling, energy demand, particle size distribution

1. Introduction

Renewable fuels have proven to enhance a clean environment, increase energy stability, and is primary in attaining sustainable development as compared to fossil-based fuels [1]. Fossil systems release higher greenhouse gases, and have adverse health and ecological impacts on the environment; and for crude oil, the geopolitics involved in the regions they mostly abound in makes it a less secure energy source [2].

Biomass waste is one of the most abundant classes of renewable fuel sources which has a lot of potential for the future energy outlook. It comes in the form of woody materials, herbaceous or aquatic plants, or manure. Woody biomass waste specifically contains fractions of cellulose, hemicellulose, lignin, and small amounts of extractives. The varying ratios of cellulose and lignin serve as a guide in identifying the suitability of biomass for energy conversion [3]. Fuels that can be obtained from woody biomass waste include ethanol, biodiesel, hydrogen, methanol, and Fischer-Tropsch diesel, among others [4].

According to Sharma et al. [5], the processes involved in converting biomass to biofuels can be grouped into two (2) broad categories;

- The thermochemical processes—this depends on heat application under controlled oxygen conditions to convert biomass to syngas, methanol or fuel oils as well as many other products. The most widely known thermochemical technologies include pyrolysis, gasification and carbonization;

- The biochemical processes—are mostly applied in converting man-made organic waste of high moisture contents and low C/H ratios, biomass is fermented to produce either biogas or organic fertilizer which are very useful products economically. The biochemical techniques also include anaerobic fermentation, methane production at landfill sites, and alcohol fermentation, although these technologies can be possible under different conditions of operation.

Considering the physical nature and sizes of the feedstock in which the woody biomass materials occur, an attempt to process them into biofuels in their naturally occurring states will be challenging. Woody biomass is relatively the largest biomass produced and therefore, there is a need to remove lignin (which has a large molecular weight and a heterogeneous structure) and also increase the cellulose surface area to facilitate their valorization into fuels [6]. Due to this, it is necessary to perform pre-treatment of biomass before it can be converted to fuels; this involves a structural alteration of the biomass to destroy its intransigent nature, increase its surface area, remove lignin, and partially hydrolyze hemicellulose [7]. In the pyrolysis of various samples of microcrystalline cellulose, xylan from beech wood and alkali lignin by Zhu and Zhong [8], the pyrolysis behavior of cellulose was observed to be fast with a reaction order of 1.38 and frequency factor of 3.5×10^{12}/s in contrast to hemicellulose and lignin due to the single unit structure of cellulose. The reaction orders for pyrolyzing lignin and hemicellulose were higher than that of cellulose at 2.30 and 1.51, and with frequency factors of 9.67×10^{9}/s and 2.59×10^{5}/s respectively. The pyrolysis of lignin also produced 47.9% coke, which is unsustainable and undesired in enhancing sustainability. Therefore, it is a critical step in biorefinery to aid improve the biomass properties for reaction and to promote efficient energy utilization in the process [9]. In this chapter, the most suitable pretreatment techniques utilized in coal-to-fuel technologies, and can be applied to suit a biomass feedstock will be further analyzed.

1.1 Comparative advantage of renewable energy to fossil fuel systems

- According to the International Energy Agency (IEA), fossil fuels accounted for approximately 89% of global CO_2 emissions [10], whereas renewable energy sources had little to no direct CO_2 emissions during their operations in 2021. For instance, the U.S. Environmental Protection Agency (EPA) reports that wind energy systems produce only 9–18 g of CO_2/kWh compared to 1 kg of CO_2/kWh for a coal-fired electricity generation process [11]. In this regard, renewable systems preserve the environment and enhance a cleaner environment as compared to their fossil counterparts.

- Bioethanol, primarily produced from corn and sugarcane, has an energy output-to-input ratio of approximately 1.3–1.6 [12], whiles gasoline from crude sources has a lower energy return on investment (EROI) of about 0.8 [13], indicating that bioenergies are more energy efficient that fossils.

- Renewable energy systems enhance energy security by diversifying the energy supply due to their abundance in almost every part of the earth, and reducing dependence on imported fuels. Fossils are however subject to price volatility and geopolitical tensions. The IEA [14] notes that renewables accounted for nearly 29% of global electricity generation in 2020, with projections to reach 55% by 2050. This shift reduces the risk of energy supply disruptions and enhances national energy independence.

- According to IRENA [15], the renewable energy industry employed over 12 million people worldwide in 2020, a figure expected to rise as the sector expands. In contrast, the fossil fuel industry faces job losses due to automation and declining demand. The transition to renewables offers opportunities for reskilling workers and fostering economic development in regions rich in renewable resources.

2. Review on the significance of biomass size reduction for use in bioenergy in reactors

In order to obtain bioenergy (biogas, biofuel of heat and electricity) from woody biomass, the lignocellulosic biomass is subjected to either of the different types of physical-mechanical pretreatment, mainly through size reduction, then by separating, before they are ultimately processed for their bioenergy products [16].

Physical pre-treatment of lignocellulosic biomass by reducing its dimensions or size is a mandatory condition prior to the chemical or bio-chemical processing in biomass bio-refining [16], for the improvement of the woody biomass property for reactions and to promote efficient energy utilization in the process [16], and to produce wood pellets or used in pulverized coal combustors.

According to Mayer-Laigle et al. [17], the targeted size to which the woody biomass is to be reduced to, is a function of the intended applications and can vary from several centimeters to a few micrometers. For example, in energetic applications, coarse particles in the range of a few millimeters (particle sizes around 0.1–0.5 mm) are targeted for anaerobic digestion processes in the production of biofuels, and below 0.1 mm for gasification processes or direct combustion in engines. Most of the time, in green chemistry applications, it is preferable to reduce the particle size below the cellular scale (~0.1 mm) to facilitate the extraction of molecules of interest.

The objective of size reduction, or comminution, is to increase the specific surface area available, thus facilitating the exchange of mass and heat and the action of anaerobic microorganisms for applications in biogas production. This is necessary to create good-flowing, easily digestible lignocellulosic biomass [18] in reaction systems. In a study by Wang et al. [19], large biomass particle sizes increase the temperature gradient in their reactions, thus affecting the local heating rate; whereas small biomass particles (<2 mm) increase the yield of bio-oils and lighter products during thermal processing, i.e., combustion, pyrolysis or gasification. Tawalbeh et al. [20] also conducted a research study based on a comprehensive scale for woody biomass pyrolysis that considered the effect of operating temperature and particle size. They discovered that an increase in moisture content led to a decrease in biomass conversion rate and the increased particle shrinkage led to an increase in biomass conversion rate in pyrolysis reactors.

Aside from increasing the accessible surface area of the woody biomass, the size reduction process also decreases the cellulose crystallinity of the feedstock, which improves its digestibility and makes the conversion of saccharides during hydrolysis more convenient [21].

In microwave pyrolysis technology, the raw biomass particle size will influence the yield of by-products of the pyrolysis process. The particle size of the sample impacts the intensity and distribution of the microwave irradiation, affecting the uniformity of the dielectric heating on the biomass. A small particle size is favorable to increasing the gaseous phase products. For example, a small size implies higher gas and bio-oil generation. Conversely, larger particle sizes will contribute to higher biochar yield [22].

In combustion studies, the particle size is also an important parameter to determine the selection of biomass feedstock to avoid fouling of combustion reactor as studied by Zaman et al. [23]. Particle size also determines the effect on the reaction time and temperature of the biomass waste. Small particles are conducive to shifting the pyrolysis process to a lower temperature, generating a quick thermochemical conversion [24].

In studying the particle size as a pyrolysis parameter, reported data from previous studies agreed with each other. Aysu [25] conducted a pyrolysis experiment at 500°C in a fixed bed reactor on four different particle sizes ranging from 0.15 to 0.85 mm. Results showed a decrease of 2.19% in oil yield as the particle size increased, with corresponding 3.96% and 5.27% increases in gas and biochar yields respectively. Pyrolysis of durian shell by Tan et al. [26] also reported no significant effects of change in particles sizes on liquid yields, but a slight increase in biochar yield. The experiment was done in a drop-type fixed bed pyrolizer at 550°C with particle size ranges from <0.5 to 5 mm. Both Aysu [25] and Tan et al. [26] concluded that the particle size effect is very little on bio-oil yield in fixed bed reactors. This is due to the fact that heat and mass transfer in the fixed bed reactors is insufficient as particle sizes increase, hence resulting in low reaction rates which will not favor an increased yield of bio-oils.

2.1 Influence of size reduction on various biofuel applications

In producing bioethanol, size reduction is essential for increasing the surface area of lignocellulosic biomass, to facilitate the enzymatic hydrolysis process. According to Li et al. [27], reducing the particle size of biomass enhances the accessibility of cellulose and hemicellulose to enzymes, thereby increasing the yield of fermentable sugars. Smaller biomass sizes also improve the efficiency of pretreatment processes, such as steam explosion and acid hydrolysis, which aid in breaking down the complex structure of lignocellulosic materials [28].

For biodiesel production, size reduction plays a significant role in the extraction of oils from oilseed crops. The process of crushing and grinding oilseeds increases the surface area, allowing for more efficient oil extraction through mechanical pressing or solvent extraction methods [29]. This is particularly important for feedstocks like jatropha and soybean, where maximizing oil yield is critical for the economic viability of biodiesel production [30].

In biogas production, size reduction is crucial for enhancing the anaerobic digestion process. Smaller biomass particles increase the surface area available for microbial attack, leading to faster and more complete degradation of organic matter [31]. This results in higher biogas yields and improved process stability. Furthermore, size reduction can help in homogenizing the feedstock, which is beneficial for maintaining consistent digestion conditions and preventing the formation of floating layers or scum in the digester [32].

3. Particle size reduction technologies for woody biomass

Lignocellulose is the basic building block of plant cells, and their stubbornness to microbial or enzymatic decay becomes a major hindrance in the quest to produce biofuels from them naturally [33] since they have the tendency to degrade the feedstock quality for further use in producing biofuels. Thus, there is a need to pretreat woody biomass to change their size, structure, and composition, in order to enhance cellulose hydrolysis to monomeric sugars which can then be processed into liquid biofuels [34]. There are a lot of physical and chemical pretreatment procedures for lignocellulosic biomass such as [35]:

- pulverization [milling],
- extrusion,
- microwaving,
- steam explosion,
- acid and alkali treatment,
- pyrolysis, among others.

The focus of this chapter will be on the application of the milling process as a means of particle size reduction. Some of the other pretreatment techniques will also be briefly discussed.

3.1 Pulverization (milling)

Milling of woody biomass is intended to reduce the particle size of biomass waste, thereby increasing its surface area with a decrease in its crystallinity [36], and is a useful technique for the further upgrade of those waste to liquid biofuels. Generally, milling processes can reduce woody biomass sizes up to 0.2 mm; nonetheless, a particle size of 0.4 mm is enough to attain the desired rate and yield of woody biomass hydrolysis.

Overall, a major setback of the milling process is its high energy demand and the capital cost of mechanical equipment. For instance, 0.5–2.15 kWh of energy is required to mill a unit mass of wood fiber between 7 and 30 min at 270 rpm. Moreover, even though those wastes are used for liquid biofuel production, the produced glucose is approximately 60% of the product content, and a xylose content close to 24% after hydrolysis is attainable [37]. Some of the types of milling processes that could be employed in particulate size reduction include hammer milling, vibratory milling, ball milling, jet milling, and knife milling, among others.

3.1.1 Hammer milling

This is a common process for reducing particle size, especially in the bioenergy and bioproducts sectors, where a system of spinning hammers is used to reduce the woody biomass to smaller particle sizes. High throughput rates, ease of use, and the variety of the types of materials that may be processed are just a few benefits that this hammer milling technology offers. Hammer mills also have the ability to minimize sizes of woody biomass to particle sizes ranging up to 0.19 mm [38].

Figure 1 gives a visual idea of how the hammer mill operates. The basic components of a hammer mill according to Singh [40], are a path for feedstock passage, hammers, a mechanism, and an auger. A route is a conduit, either vertical or horizontal, that contains the object being crushed. The material inside the route is crushed using hanging hammers to produce a hammer meal. The hammers that are used to smash the material into powder are also rotated by pinions and gears. In essence, a high-speed rotor rotating inside a cylindrical container makes up a hammer mill. The shaft on which the rotor is placed is typically horizontal. A rotor disk is pinned to the swing hammers. The hammers are usually made of straight metal bars with expanded or plain ends. The swing hammer sets in this mill and breaks the particles into smaller pieces. The lower part of the enclosure is made up of a grate or screen through which the product drops. On a single shaft, several rotor disks with four to eight swing hammers apiece are frequently

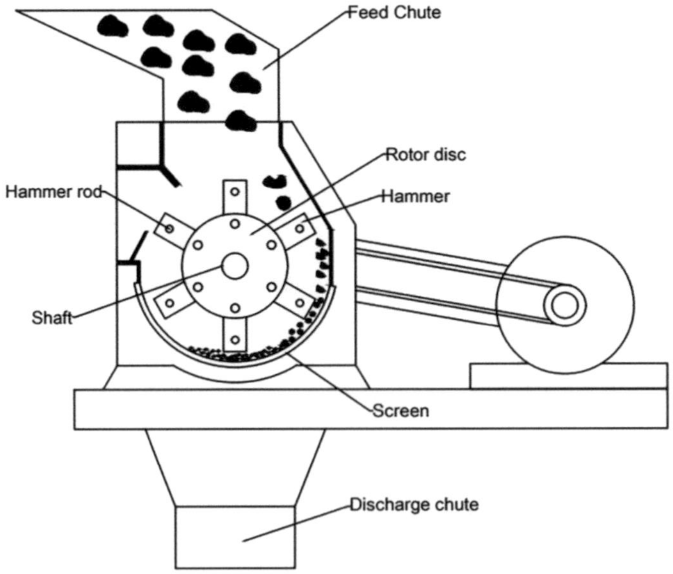

Figure 1.
An illustration of the hammer milling process [39].

installed. The diameter of the rotor disk varies from 150 to 250 mm. The hammers are hinged; thus, the presence of any hard substance does not harm the apparatus. When the hammers become worn out, they are easily replaced.

Particle sizes between 25 and 0.07 mm can be produced using intermediate hammer mills. In a hammer mill used for fine manufacturing, the hammer tips' peripheral speed may reach about 112 m/s and they grind the feed materials at a rate of 0.1–15 tons/h [40] to sizes finer than 0.07 mm.

Hammer mills are also appropriate for industrial-scale bioenergy operations because of their design and setup, which enables them to process huge amounts of raw material quickly. Therefore, the high throughput rate of up to 15 tons/h in the hammer milling for woody biomass is one of its main advantages [41]. This promotes a high production rate and shortens milling duration, increasing its overall output. Another benefit of hammer mills is how simple they are to use. They frequently have straightforward designs and few moving components, making them simple to use and maintain [42] due to less wear and tear occurrence. Furthermore, thanks to their automated feeding systems, hammer mills can operate with little operator involvement. The workflow is streamlined, and less manpower is needed.

3.1.2 Knife milling

Knife milling is a method for reducing particle size that is frequently used to process woody biomass. In this process, the biomass waste is cut and shredded into smaller pieces using sharp knives. The key benefit of knife mills is their capacity to generate a wide range of particle sizes, which makes it appropriate for a variety of uses like fuel production and chemical extraction (**Figure 2**) [43].

Knife milling operates on a straightforward principle: the woody biomass is fed into the chamber, where it is sheared and sliced into smaller pieces by rotating

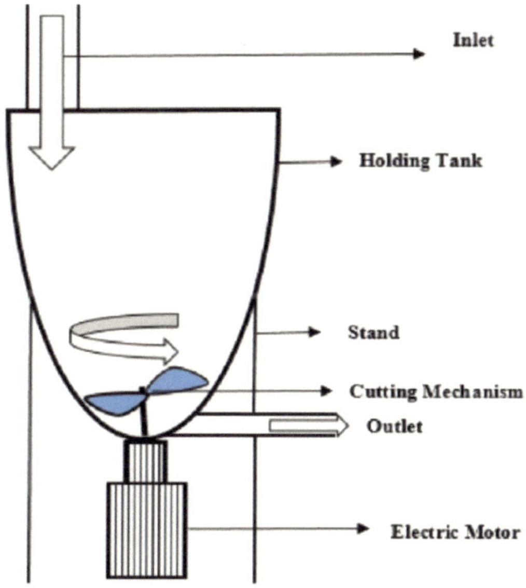

Figure 2.
An illustration of the knife milling process [44].

blades placed on a rotor. By altering the rotor's speed, the number and placement of the knives, and the size of the screen that the particles travel through, the size of the particles may be regulated [45]. Knife milling is a more energy-efficient technique than others because it uses mechanical forces rather than heat or chemicals to break down the biomass. About 19.3–32.5 kWh/ton of energy is consumed for the milling of fresh woody biomass waste while 17.8–23.2 kWh/ton is utilized for dry woody biomass waste. Also, the milling of dry woody biomass waste reduces the milling energy requirement by about 38% as compared to a wet feedstock [46].

3.1.3 Ball milling

Due to its potential applications in numerous industries, including the manufacture of biofuel, ball milling is a particle size reduction technology that has attracted a lot of attention recently. Particle sizes produced by ball milling range from 1 to 30 mm. One of ball milling's key benefits is its capacity to effectively reduce the particle size of woody biomass, which is essential for boosting the generation of ethanol from lignocellulosic materials. The procedure uses revolving drums or vessels with grinding balls to force the biomass particles to collide and fragment into smaller pieces. Additionally, Sitotaw et al. [47] discovered that using a ball milling process in conjunction with other chemical and physicochemical pretreatment methods makes it easier to remove lignin, reduce cellulose crystallinity, and increase specific surface area, all of which improve the digestibility of lignocellulosic biomass when used as a feedstock to produce liquid biofuels. Ball mills destroy the cell walls of some woods and remove their crystalline structure [48], which is a significant drawback of the process. However, Yuan et al. [49] also found out that they can also break down the fibrous structure of lignocellulosic biomass, and this is favorable to use the biomass particles in pulverized combustors or pelletize them before they are introduced into the combustors to generate energy.

According to a study by Perov [50], an industrial ball mill is made up of a steel drum that is cylindrical and contains steel balls. The cylindrical steel drum is between 3 and 15 m long and 2–5 m in diameter. The diameter of the steel balls can reach about 25 cm. The circular steel drum uses up to 1 MW of electricity to power itself. A feeder system supplies the raw material to be ground into the steel drum's inlet, and a ventilation system moves the ground materials out of the drum.

The lifting and tumbling action is what causes the crushing and grinding; hence the rotation speed of the balls is selected to best suit this operation. Spillage is avoided by installing a steel wave-shaped lining inside the inner case of the mill, and it also creates a nice tumbling movement, this helps keep the revolving shell to the charge [51].

Grinding mills are significant energy consumers because grinding requires a lot of power, according to Levit [52].

The loading of a ball mill determines how productively it grinds. The mill grinds more efficiently as more material is fed into it. When the feedstock is oversaturated in the mill, grinding productivity declines, and a mill blockage happens. This might be caused by the characteristics of the dusty system and the properties of the material being ground. Operators purposely reduce the output of the mill's grinding by adding less material, which increases the amount of power required and decreases the effectiveness of the operation [52]. **Figure 3** shows the inner cross-sectional view of a ball mill during operation.

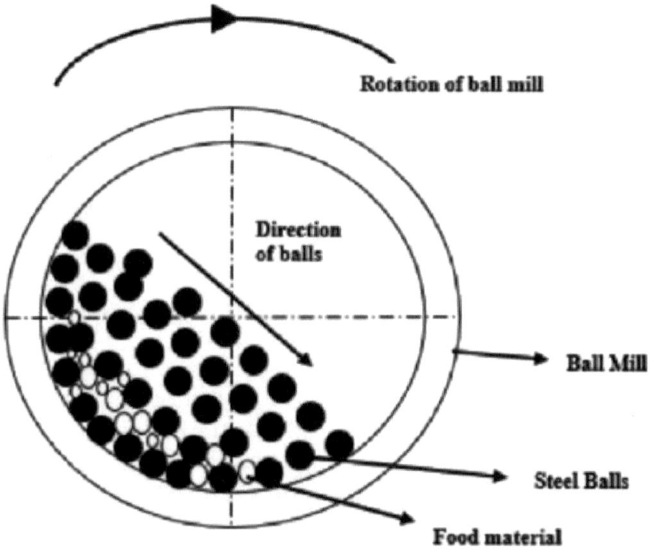

Figure 3.
A cross-sectional view of the ball mill's chamber in operation [43].

3.2 Other pretreatment methods

3.2.1 Extrusion

The extrusion process involves the application of heat (beyond 300°C) to the lignocellulosic biomass under shear stress produced by rotating screw blades in a barrel-like chamber. The combination of heat and shear forces results in the disruption of the crystalline cellulose structure hence leading to a particle size reduction [53]. The major hindrance in the industrial commercialization of this technology is the relatively high energy it consumes as compared to the other pretreatment methods. The energy required can reach as high as 0.324 kWh/kg [37] of woody biomass and is therefore not the most desired due to this economic challenge.

3.2.2 Steam explosion

This method involves the subjection of woody biomass waste to high-pressure saturated steam of between 0.7 and 4.8 MPa at average temperatures of around 200°C [54], and then a sudden depressurization step, making the biomass waste undergo an explosive decompression. This results in a degradation of the hemicellulose and a transformation of the lignin due to the temperatures involved, hence increasing the potential of cellulose hydrolysis [34] for further use.

3.3 Comparison of the three (3) milling processes discussed

To efficiently trim down the size of woody biomass to at most 0.4 mm [37] for liquid biofuel production, it is of interest to select the most efficient milling technology.

Considering the hammer milling technology, it is a highly adaptable and flexible process. It can be used to handle a variety of woody biomass feedstocks such as wood chips, sawdust, bark, and even complete logs [38]. Its flexibility in processing various feedstock types is very useful for a heterogeneous woody biomass feedstock. Hammer mills can also be easily integrated in any existing industrial process or designed to meet any specific needs.

However, there are significant drawbacks to hammer milling for woody biomass despite these benefits. According to Bitra et al. [42], hammer mills often generate products with inconsistent sizes due to varying moisture content, feedstock characteristics, and operating conditions. Getting a uniform particle size distribution therefore becomes difficult due to the nature of the impact forces between the hammers and the biomass particles that cause the size reduction. The presence of moisture in woody biomass is also another restriction for its usage in hammer mills. Serrano et al. [55] indicate that a high moisture content can reduce throughput rates and increase energy consumption in hammer mills. This problem can be solved by drying the woody biomass before milling or by including a drying step in the procedure. The potential for excessive energy consumption is still another drawback for effective size reduction in hammer mills; and this may lead to high operational costs when milling woody biomass. To reduce energy demand while maintaining appropriate particle sizes, it is essential to adjust operating parameters such as hammer speed and screen size [42] and these may sometimes be counter-productive, hence negatively impacting operational costs.

The knife milling technique is suited for a variety of applications due to its flexibility. Its adaptability in producing a range of particle sizes is one of its main benefits. It is possible to create a wide range of particle sizes, from coarse chips to fine powders, by modifying the blade sizes or diameters [42]. While finer particles are preferable for chemical extraction operations like biofuel production or pharmaceutical manufacture, relatively larger particles can be employed for fuel production in boilers or pellet mills [56]. Knife mills also have a lower initial investment cost than other size-reduction technologies and as compared with hammer and ball mills; they are less expensive to acquire and maintain [21]. Due to its direct cutting action, knife milling also requires less energy, which reduces operational costs. Knife milling is a desirable alternative for small-scale wood biomass processing enterprises or research laboratories with tight resources because of its affordability [57].

Knife milling does have some challenges as well. Due to its relatively low throughput and vulnerability to damage from hard contaminants like rocks and pebbles, which may be picked up in previous biomass operations, knife mills may be unfavorable to use for impure feeds [58].

The ball milling process also gives room for the incorporation of chemical pretreatment, with a possible minimum particle size of 0.009 mm attainable [59]. It can also handle any degrees of moisture content in the woody biomass, even though this can increase the specific energy demand to the highest as compared to the other milling processes. Therefore, knife mills can be deemed as the most suitable and efficient industrial milling processes since aside all its merits, they can handle high feedstock throughput with about 25 wt.% moisture content with minimum specific energy demands [21]. Ball milling is however a mostly used technique for laboratory-scale processes (**Table 1**) [59].

3.4 Choice of best-suited particle size reduction technology of woody biomass

It has prior been established by Kratky and Jirout [21], and Arce and Kratky [59] that the knife milling process is the industrially most preferred milling process for woody biomass waste. Some of the factors that affect power consumption and overall

Type of mill	Energy consumption
Hammer mill	≤60 kWh/ton [46, 60]
Knife mill	≤32.5 kWh/ton [46, 60]
Ball mill	≤2.15 MWh/ton [46]

Table 1.
Comparison of hammer, knife and ball mills.

performance in mills include the hardness of mill feedstock, fineness of the milled product, circulating load and milling mechanism, and the number of milling stages [61].

Knife mills are able to handle hard feed, ranging from coals, dry wood, among others. The shearing and cutting technique used by the knife blades in size reduction at their speeds of average 1500 rpm makes them able to cut through hard feed materials and get them to the desired sizes with relatively less wear or power consumption as compared to hammer mills and ball mills which will take longer to achieve such fineness. Knife mills are also able to generate milled products of high consistency in particle size distribution due to the uniformity in the milling process as compared to hammer and ball mills.

Another important point to note is the fact that high wear-resistant blades can be applied in knife mills. Grejtak et al. [62] experimentally deduced that tungsten-carbide (WC-Co) or iron-bonded blades generated about thrice the milling throughput as with the conventional steel blades, with the normalized power consumption and no need for increased power demand. This can also be a contributing factor to a higher production rate of milled products in the knife mills when compared to the other milling processes. The use of these more powerful blades can reduce the milling downtime by about 6–85% with a $2–3 per ton cut in the cost of milling. This is supported by Kratky and Jirout [21] when they compared the cost of the knife milling process to the other size reduction technologies. It also confirms the assertion by Williams et al. [57] about the reduced energy requirement in the knife milling process which reflects in comparatively low operational expenditure.

Ball mills and hammer mills fall short of the possible milling results achievable in knife mills. Hammer mills have a simple design and are a good process choice for their ruggedness and versatility in milling feedstock [63]. Despite these positives, they are not capable of reducing the sizes of moisturized woody biomass waste as compared to knife mills. Knife mills are also about 25% more energy-efficient than hammer mills making them less expensive.

Another downside of hammer mills is their inability to produce a consistent grind profile of milled biomass waste as compared to knife mills. Due to the nature of the milling process in knife mills, there is a greater consistency in the particle size distribution of the product output [64]. The bioenergy production is hugely dependent on the woody biomass waste particle size distribution as further explained in a study by Su and Yu [65]. Using the Rosin-Rammler (RR) distribution to assess the particle size distribution in knife mills deduces that the RR distribution range for knife mills is narrow as compared to hammer mills. For instance, in the application of this distribution test to corn stover, the geometric mean of particles after knife milling was 0.541 mm, 0.626 mm, and 0.814 mm, and they were 0.621 mm, 0.959 mm, and 1.764 mm for hammer mills at the same screen sizes of 3.2 mm, 6.3 mm, and 12.7 mm respectively. This observed result will also be experienced in a similar analysis for woody biomass waste, and it gives a very clear understanding of the particle size distribution of milling in both knife and hammer mills, and makes knife mills much more efficient to use than hammer mills to achieve a consistent product size distribution. With an increase in the screen size from 3.2 to 12.7 mm in the experiment by Su and Yu [65], the Rosin-Rammler (RR) distribution parameter decreased from 2.113 to 1.448 for knife mill and from 1.951 to 1.397 for hammer mill. RR distribution parameter represented the uniformity of particles, hence a decreased value of RR distribution parameter represented less uniformity of particles. The knife mill has a higher value of RR distribution parameter than the hammer mill at the same screen size levels, indicating a better product size consistency for knife mills.

For ball mills, although they are noted for being compatible with other chemical pretreatment methods to attain desired products with minimized effort it still comes up short when compared to knife mills. Smaller grinding balls (about 3 mm) and longer durations are required to produce finer particles in ball mills. A significant amount of energy is also required due to the gradual wearing of the balls over time, friction between the milling materials and balls, and the subsequent gradual heating of the mill materials [66]. Although the blades of knife mills can also undergo wearing, they are far less in number as compared to the number of balls in a ball mill and can be easily replaced as compared to ball mills. Ball mills are also generally large and very heavy making their handling less convenient. It is also easier to replace individual hammers or blades in hammer and knife mills respectively as compared to balls in ball mills due to their sizes; it may be difficult to identify the balls with defaults unless all the balls are taken out and carefully observed. In the milling of mixed wood pellets by Williams et al. [67], the milling energy was expressed as a percentage of the dry high heating value of the mixed feed sample. The energy requirement for ball mills represented a 66% fraction whiles that of a knife mill was around 57%, and this indicates how knife mills are more energy efficient over ball mills.

3.5 Impact of biomass particle size on the choice of reactor

It is important to analyze the factors that affect the efficient feedstock feeding in fixed and fluidized bed reactors in addition to the bioenergy generation from these reactors. Some of the critical factors include: reaction temperature, biomass type, biomass particle size, biomass composition and moisture level, and reactor efficiency, among others [68]. The impact of biomass particle size on the reactor choice will be explored further.

He et al. [69] observed a phenomenon during the torrefaction of Paulownia wood in a fixed bed reactor. Upon reducing the feedstock size from 12 to 0.3 mm, the solid product yield reduced by 5.41 wt.% at 260°C and 3.54 wt.% at 290°C; however, the liquid product yield increased by 5.87 wt.% at 260°C and 3.25 wt.% at 290°C, and the variation in gaseous product yield was negligible.

Agu et al. [70] assessed the behavior of biomass wood pellets and chips of different sizes (8.96 mm and 6.87 mm respectively) in a fluidized bed reactor. The results indicated that the wood chips (smaller sized particles) separated upwardly whereas the wood pellets (larger sized particles) separated downwardly, with the upward separation of the chips increasing with higher gas velocities. According to Narnaware et al. [71], reducing biomass size from 12 to 0.3 mm increases the yield of gaseous products for the production of syngas (H_2, CO_2, CO, and CH_4 traces) in a fluidized bed reactor.

Yaghoubi et al. [72] also studied the effects of varying process parameters on H_2–rich syngas production through gasification in a fluidized bed reactor. In the study,

Particle size (mm)	Mole fractions			
	H_2	CO_2	CO	CH_4
200	0.315	0.205	0.355	0.125
400	0.310	0.200	0.365	0.125
600	0.300	0.195	0.375	0.130
800	0.295	0.195	0.380	0.130
1000	0.290	0.195	0.385	0.130

Table 2.
Mole fractions of syngas components trend against varying woody biomass sizes in a fluidized bed reactor [72].

the hydrodynamics of the bed is hugely dependent on the particle size of woody biomass waste. For heavy particle sizes, the minimum fluidization velocity will approach the inlet gas fluidization velocity and this makes the hydrodynamic condition of the particle bed lean towards minimum fluidization conditions. Due to this, H_2 production reduces with increasing woody biomass particle size and vice versa. **Table 2** shows the trend of product composition in the syngas for this study at varying woody biomass particle sizes.

In a comparison of olive kernel biomass pyrolysis in both thermogravimetric fixed and fluidized bed reactors [73], the particle size had no significant effect on the rate of reaction in the fixed bed; whereas the reaction duration remained fairly constant in both reactors for reaction temperatures less than 450°C. For reacting temperatures above 500°C in the fluidized bed, the reaction duration increased with an increase in particle size, thereby slowing the reaction process. This phenomenon in the fluidized bed was also observed by Heidari et al. [74] when analyzing the effect of reacting conditions on product yield for pyrolyzing *Eucalyptus grandis* in a fluidized bed reactor. Larger particle sizes were found to lead to an increase in char production and a decrease in bio-oil production due to the slow heating of particles. Hence smaller biomass sizes (6 mm or less) are able to achieve a higher biomass heating rate in the fluidized bed reactor. For a particle size of 1.5 mm, a 61% bio-oil yield was attained with a char yield of 13.6% and a gas yield of 25.3%. However, for a particle size of 3.5 mm, the bio-oil, char, and gas yields were 54%, 16.9%, and 28.3% respectively.

For a desired liquid fuel product during fluidized bed pyrolysis of biomass waste, particle size ranges of 0.1–6 mm are appropriate at temperatures between 425 and 525°C; for commercial plants, 2–5 mm can also yield the desired product compositions after pyrolysis [75].

From these findings, it can be observed that size reduction in fluidized beds results in a H_2 – rich stream of syngas product which is highly desired in the synthesis of bioenergy products. Bio-oil yields also experienced an increase in yield when pyrolized in fluidized bed by about 7%. There also are some energy savings for pyrolyzing reduced particle sized in fluidized beds due to the about 3.3% drop in biochar yield. It also translates to a reduction in the pollution effects of the pyrolysis product obtained. The process also generates an overall useful bioenergy product which constitutes 86% of the product stream and is a very reliable bioenergy source [74]. The biochar itself can have other uses in the energy value chain, even though its production is not very much desired. In fixed beds however, the effect of particle size variations on heat and mass transfer is negligible, hence reducing particle sizes does not have any significant effect on producing more quantities of the desired bioenergy products. Factors such as temperature and catalyst type rather have major effects on biomass conversion to bioenergy products in fixed beds [25].

3.6 Environmental impact of milling processes

The environmental impacts of milling primarily stem from its energy consumption and the potential release of particulate matter. Milling is an energy-intensive process, and the type of mill used significantly influences the energy requirements. High energy consumption not only increases operational costs but also contributes to greenhouse gas emissions if the energy is sourced from fossil fuels. Therefore, optimizing milling conditions to reduce energy use is crucial for minimizing its environmental footprint [76]. Milling can also generate dust and particulate matter, which can pose health risks to workers and contribute to air pollution if not properly managed. Implementing dust control measures, such as using enclosed milling systems and installing air filtration units, can mitigate these impacts [28].

3.7 Scalability of milling processes for industrial applications

Scalability is a significant consideration for the industrial application of milling as a biomass pretreatment process. One of the main challenges is maintaining efficiency and cost-effectiveness when transitioning from laboratory-scale to industrial-scale operations. The scalability of milling is influenced by factors such as equipment design, energy consumption, and throughput capacity.

To achieve scalability, it is essential to select the appropriate milling technology that can handle large volumes of biomass while maintaining consistent particle size reduction. Advances in milling technology, such as the development of more energy-efficient mills and the integration of milling with other pretreatment processes, can enhance scalability [77]. Pretreatment methods that can be integrated can be chemical or biological treatments, which aid in improving the overall efficiency of biomass conversion. This integrated approach can reduce the energy requirements of milling and enhance the digestibility of biomass, making it more feasible for large scale applications [78].

4. Summary

This chapter is aimed at assessing the size reduction of woody biomass waste, to sizes that will be useful for bioenergy production in fixed or fluidized bed reactors. The various size reduction techniques were analyzed to choose the most suitable process to achieve the desired product particle sizes in a cost-effective, energy-efficient, and environmentally friendly manner. Among the three milling techniques scrutinized, that is the hammer milling, ball milling, and knife milling processes, the knife milling technology was chosen as the most suitable for the reduction of the sizes of woody biomass waste materials. The knife mill affords flexibility in the range of product particle sizes since the sizes and diameters of the cutting blades can be adjusted, and they handle materials with moisture better than the other mill types. They are also relatively cheaper as compared to hammer and ball mills with less energy requirements for milling equal quantities of the same materials. Also, knife mills are able to produce the desired size of woody biomass waste for conversion to biofuels in fixed and fluidized beds. Even though biomass sizes of 0.30 mm will be enough for biofuel production in reactors, the knife mill is capable of achieving 0.25 mm minimum size distribution of products, and hence a very good choice.

For industrial applications, drying woody biomass before comminution will positively ehance the economics of the process. The relatively high energy consumption in the size reduction of fresh waste wood chips is most likely owing to the moisture in the biomass acting as a plasticizer. Drying the wood chips before milling can reduce the energy requirement by 38% and this will translate in very significant cost savings for industry. The use of sawdust as an industrial feed source is also worth assessing. In a study by Tryjarski et al. [60], the energy required to reduce the size of wood shavings of pine wood feedstock of 7% moisture content was just the same as for a woody biomass with 10% moisture content. However, the lesser moisture content in the wood shavings will translate in a higher efficiency, and this will be interesting to further explore for industrial applications.

Author details

Maxwell Appiah[1*] and Prince L. Tetteh[2]

1 University of Hull, England

2 Kwame Nkrumah University of Science and Technology, Ghana

*Address all correspondence to: appiahmaxwell978@gmail.com

References

[1] Petrović-Ranđelović M, Kocić N, Stojanović-Ranđelović B. The importance of renewable energy sources for sustainable development. Economics of Sustainable Development. 2020;**4**(2):15-24

[2] Boundy B, Diegel SW, Wright L, Davis S. Biomass Energy Data Book. Lincoln: US Department of Energy Publications; 2011

[3] McKendry P. Energy production from biomass (part 1): Overview of biomass. Bioresource Technology. 2001;**83**:37-46

[4] Wright L, Boundy B, Perlack B, Davis S, Saulsbury B. Biomass Energy Data Book. Lincoln: US Department of Energy Publications; 2006

[5] Sharma S, Meena R, Sharma A, Goyal PK. Biomass conversion technologies for renewable energy and fuels: A review note. IOSR Journal of Mechanical and Civil Engineering. 2014;**11**(2):28-35

[6] Ning P, Yang G, Hu L, Sun J, Shi L, Zhou Y, et al. Recent advances in valorization of plant biomass. Biotechnology for Biofuels. 2021;**14**(102):1-22

[7] Chandel H, Kumar P, Chandel AK, Verma ML. Biotechnological advances in biomass pretreatment for bio-renewable production through nanotechnological intervention. Biomass Conversion and Biorefinery. 2022;**14**(3):2959-2981. DOI: 10.1007/s13399-022-02746-0

[8] Zhu L, Zhong Z. Effects of cellulose, hemicellulose and lignin on biomass pyrolysis kinetics. Korean Journal of Chemical Engineering. 2020;**37**(10):1660-1668

[9] Anukam A, Berghel J. Biomass pretreatment and characterization: A review. Biotechnological Applications of Biomass. 2020:1-17. DOI: 10.5772/INTECHOPEN.93607

[10] IEA. Global Energy Review: CO_2 Emissions in 2021. International Energy Agency; 2022. Available from: https://www.iea.org/reports/global-energy-review-co2-emissions-in-2021-2

[11] EPA. US Environmental Protection Agency. Greenhouse Gas Emissions from a Typical Passenger Vehicle. 2021. Available from: https://www.epa.gov/greenvehicles/greenhouse-gas-emissions-typical-passenger-vehicle

[12] Hill J et al. Environmental, economic, and energetic costs and benefits of biodiesel and ethanol biofuels. Proceedings of the National Academy of Sciences. 2006;**103**(30):11206-11210

[13] Murphy DJ et al. Order from chaos: A preliminary protocol for determining the EROI of fuels. Sustainability. 2011;**3**(10):1888-1907

[14] IEA. Net Zero by 2050: A Road Map for the Global Energy Sector. International Energy Agency; 2021. Available from: https://www.iea.org/reports/global-energy-review-co2-emissions-in-2021

[15] IRENA. Renewable Energy and Jobs: Annual Review 2021. International Renewable Energy Agency; 2021. Available from: https://www.irena.org/publications/2021/Oct/Renewable-Energy-and-Jobs-Annual-Review-2021#:~:text=This%20most%20recent%20report%20estimates,estimated%207.3%20million%20in%202012

[16] Moiceanu G, Paraschiv G, Voicu G, Dinca M, Negoita O, Chitoiu M, et al. Energy consumption at size reduction of lignocellulose biomass for bioenergy. Sustainability. 2019;**11**(9):2477

[17] Mayer-Laigle C, Blanc N, Rajaonarivony RK, Rouau X. Comminution of dry lignocellulosic biomass, a review: Part I – From fundamental mechanisms to milling behaviour. Bioengineering. 2018;**5**(2):41

[18] Oyedeji O, Gitman P, Qu J, Webb E. Understanding the impact of lignocellulosic biomass variability on the size reduction process: A review. ACS Sustainable Chemistry and Engineering. 2020;**8**(6):2327-2343

[19] Wang Y, Akbarzadeh A, Chong L, Du J, Tahir N, Awasthi MK. Catalytic pyrolysis of lignocellulosic biomass for bio-oil production: A review. Chemosphere. 2022;**297**:134-181

[20] Tawalbeh M, Al-Othman A, Salamah T, Alkasrawi M, Martis R, El-Rub ZA. A critical review on metal-based catalysts used in the pyrolysis of lignocellulosic biomass materials. Journal of Environmental Management. 2021;**299**:113597

[21] Kratky L, Jirout T. Biomass size reduction machines for enhancing biogas production. Chemical Engineering & Technology. 2011;**34**(3):321-487

[22] Allende S, Brodie G, Jacob MV. Breakdown of biomass for energy applications using microwave pyrolysis: A technological review. Environmental Research. 2023;**226**:115619

[23] Zaman KK, Balasundram V, Ibrahim N, Samsudin MDM, Kasmani R, Hamid MKA, et al. Effect of particle size and temperature on pyrolysis of palm kernel shell. International Journal of Engineering and Technology (UAE). 2018;**7**(4):118-124

[24] Leng L, Xiong Q, Yang L, Li H, Zhou Y, Zhang W, et al. An overview on engineering the surface area and porosity of biochar. Science of the Total Environment. 2021;**763**:144204

[25] Aysu T. Catalytic pyrolysis of eremurus spectabilis for bio-oil production in a fixed-bed reactor: Effects of pyrolysis parameters on product yields and character. Fuel Processing Technology. 2015;**129**:24-38

[26] Tan YL, Abdullah AZ, Hameed BH. Fast pyrolysis of durian (*Durio zibethinus* L) shell in a drop-type fixed bed reactor: Pyrolysis behaviour and product analyses. Bioresource Technology. 2017;**243**:85-92

[27] Li X et al. Improving enzymatic hydrolysis of lignocellulosic biomass by bio-coordinated physicochemical pretreatment. Fermentation. 2022;**8**(10):558

[28] Sun Y, Cheng J. Hydrolysis of lignocellulosic materials for ethanol production: A review. Bioresource Technology. 2002;**83**(1):1-11

[29] Knothe G, Gerpen JV, Krahl J. The Biodiesel Handbook. s.l. Champaign, Illinois: AOCS Press; 2005. pp. 12-19

[30] Gui MM, Lee KT, Bhatia S. Feasibility of edible oil vs. non-edible oil vs. waste edible oil as biodiesel feedstock. 2008;**33**(11):1646-1653

[31] Angelidaki I, Ellegaard L, Ahring BK. Applications of the anaerobic digestion process biomethanation II. Advances in Biochemical Engineering/Biotechnology. 2003;**82**:1-33

[32] Weiland P. Biogas production: Current state and perspectives. Applied

Microbiology and Biotechnology. 2010;**85**(4):849-860

[33] Zoghlami A, Paës G. Lignocellulosic biomass: Understanding recalcitrance and predicting hydrolysis. Frontiers in Chemistry. 2019;**7**(874):1-11

[34] Kumar P, Barrett DM, Delwiche MJ, Stroeve P. Methods for pretreatment of lignocellulosic biomass for efficient hydrolysis and biofuel production. Industrial & Engineering Chemistry Research. 2009;**48**:3713-3729. DOI: 10.1021/ie801542g

[35] Cheah WY, Sankaran R, Show PL, Tg Ibrahim TNB, Chew KW, Culaba A, et al. Pretreatment methods for lignocellulosic biofuels production: Current advances, challenges and future prospects. Biofuel Research Journal. 2020;**25**:1115-1127

[36] da Silva AS, Teixeira RSS, Moutta RO, Ferreira-Leitão VS, de Barros RRO, Ferrara MA, et al. Sugarcane and Woody Biomass Pretreatments for Ethanol Production. InTech, Editors: Anuj K. Chandel, Silvio Silvério da Silva. 2013. DOI: 10.5772/53378

[37] Baruah J, Nath BK, Sharma R, Kumar S, Deka RC, Baruah DC, et al. Recent trends in the pretreatment of lignocellulosic biomass for value-added products. Frontiers in Energy Research. 2018;**6**(141):1-19

[38] Mani S, Tabil LG, Sokhansanj S. Grinding performance and physical properties of wheat and barley straws, corn stover, and switchgrass. Biomass & Bioenergy. 2004;**27**(4):339-352

[39] Nitesh. What is a Hammer Mill Machine? 2023. Available from: https://karkhana.io/blog/hammer-mill-machine/

[40] Singh B. Hammer Mill Construction, and Working Principal, Chemicals Learning. 2021. Available from: https://www.chemicalslearning.com/2022/11/hammer-mill-construction-and-wroking.html?expand_article=1

[41] Tumuluru JS, Tabil LG, Song Y, Iroba KL, Meda V. Grinding energy and physical properties of chopped and hammer-milled barley, wheat, oat, and canola straws. Biomass & Bioenergy. 2014;**60**:58-67

[42] Bitra VSP, Womac AR, Chevanan N, Miu PI, Sokhansanj S, Igathinathane C. Direct mechanical energy measures of hammer mill comminution of switchgrass, wheat straw, corn stover and analysis of their particle size distribution. Powder Technology. 2009;**193**(1):32-45

[43] Kaur G, Orsat V, Singh A. An overview of different homogenizers, their working mechanisms and impact on processing of fruits and vegetables. Critical Reviews in Food Science and Nutrition. 2021;**63**(14):1-14

[44] Paulrud S, Mattsson JE, Nilsson C. Particle and handling characteristics of wood fuel powder: Effects of different mills. Fuel Processing Technology. 2002;**76**(1):23-39

[45] Miao Z, Grift TE, Ting KC. Size Reduction and Densification of Lignocellulosic Biomass Feedstock for Biopower, Bioproducts, and Liquid Biofuel Production. 2010. Available from: http://abe-research.illinois.edu/pubs/T_Grift/E-EAFE2-120051298.pdf

[46] Paraschiv G, Moiceanu G, Voicu G, Chitoiu M, Cardei P, Dinca MN, et al. Optimization issues of a hammer mill working process using statistical modelling. Sustainability. 2021;**13**(973):3

[47] Sitotaw YW, Habtu NG, Gebreyohannes AY, Nunes SP, Gerven TV. Ball milling as an important pretreatment technique in lignocellulose biorefineries: A review. Biomass Conversion and Biorefinery. 2021;**13**:15593-15616

[48] Agarwal UP, Zhu J, Ralph SA. Enzymatic hydrolysis of loblolly pine: Effects of cellulose crystallinity and delignification. Holzforschung. 2013;**67**(4):371-377

[49] Yuan X, Liu S, Feng G, Liu Y, Li Y, Lu H, et al. Effects of ball milling on structural changes and hydrolysis of lignocellulosic biomass in liquid hot-water compressed carbon dioxide. Korean Journal of Chemical Engineering. 2016;**33**(7):2134-2141

[50] Perov V, Andreev E, Bilenko L. Grinding, Milling and Screening of Mineral Products. Moscow: Nedra Publishing House; 1990

[51] Luckie PT, Austin LG. Coal Grinding Technology: A Manual for Process Engineers. US Department of Energy; 1990. pp. 16-17

[52] Levit G. Production of Dust at Thermoelectric Power Stations. Moscow: Energoatomizdat Publishing House; 1991 5-283-00151-2

[53] Kumar AK, Sharma S. Recent updates on different methods of pretreatment of lignocellulosic feedstocks: A review. Bioresources and Bioprocessing. 2017;**4**(7):1-19

[54] Hernández-Beltrán JU, Hernández-de LIO, Cruz-Santos MM, Saucedo-Luevanos A, Hernández-Terán F, Balagurusamy N. Review: Insight into pretreatment methods of lignocellulosic biomass to increase biogas yield: Current state, challenges, and opportunities. Applied Sciences. 2019;**9**(3721):1-29

[55] Serrano C, Monedero E, Lapuerta M, Portero H. Effect of moisture content, particle size and pine addition on quality parameters of barley straw pellets. Fuel Processing Technology. 2011;**92**(3):699-706

[56] Mahapatra S, Kumar D, Singh B, Sachan PK. Biofuels and their sources of production: A review on cleaner sustainable alternative against conventional fuel, in the framework of the food and energy nexus. Energy Nexus. 2021;**4**(100036):1-12

[57] Williams O, Lester E, Kingman S, Giddings D, Lormor S, Eastwick C. Benefits of dry comminution of biomass pellets in a knife mill. Biosystems Engineering. 2017;**160**:42-54

[58] Westover TL, Hartley D. Biomass Handling and Feeding. In Book: Advances in Biofuels and Bioenergy. 2018. DOI: 10.5772/intechopen.74606

[59] Arce C, Kratky L. Review: Mechanical pretreatment of lignocellulosic biomass toward enzymatic/fermentative valorization. iScience. 2022;**25**(7):1-25

[60] Trykarski P et al. Physicomechanical properties of raw and comminuted pine and poplar shavings: Energy consumption, particle size distribution and flow properties. Wood Science and Technology. 2023;**57**:1-25

[61] Jorge. Factors Having an Effect on the Power Consumption and Capacity of a Mill. 2019. Available from: https://www.911metallurgist.com/factors-having-an-effect-on-the-power-consumption-capacity-of-a-mill/

[62] Grejtak T, Lacey JA, Kuns MW, Damon SH, Thompson DN, Fenske G, et al. Improving knife milling performance for biomass preprocessing

by using advanced blade materials. Wear. 2023;**522**(204714):1

[63] Yancey N, Wright CT, Westover TL. Optimizing hammer mill performance through screen selection and hammer design. Biofuels. 2013;**4**(1):85-94

[64] RMS. Does a Roller Mill have more Advantages than a Hammer Mill? Read This! 2022. Available from: https://rmsroller-grinder.com/blog/roller-mill-vs-hammer-mill/

[65] Su D, Yu M. Study of corn stover particle size distribution characteristics for knife mill and hammer mill. IOP Conference Series: Earth and Environmental Science. 2019;**358**(052060):1-7

[66] Neikov OD. Chapter 2 – Mechanical crushing and Grinding. In: Handbook of Non-Ferrous Metal Powders: Technologies and Applications. Kiev: s.n.; 2009. pp. 47-62

[67] Williams O et al. Influence of mill type on densified biomass comminution. 2016;**182**:219-231

[68] Aguilar G, Muley PD, Henkel C, Boldor D. Effects of biomass particle size on yield and composition of pyrolysis bio-oil derived from Chinese tallow tree (*Triadica Sebifera* L.) and energy cane (*Saccharum complex*) in an inductively heated reactor. AIMS Energy. 2015;**3**(4):838-850

[69] He Y, Zhang S, Liu D, Xie X, Li B. Effect of biomass particle size on the torrefaction characteristics in a fixed-bed reactor. Energies. 2023;**16**(1104):1-14

[70] Agu CE, Tokheim LA, Pfeifer C, Moldestad BME. Behaviour of biomass particles in a bubbling fluidized bed: A comparison between wood pellets and wood chips. Chemical Engineering Journal. 2019;**363**:84-98

[71] Narnaware SL, Panwar NL, Gupta T, Meena KK. Bubbling fluidized bed gasification of biomass: A review on the effect of selected operational parameters. Biointerface Research in Applied Chemistry. 2022;**13**(5):1-32

[72] Yaghoubi E, Xiong Q, Doranehgard MH, Yeganeh MM, Shahriari G, Bidabadi M. The effect of different operational parameters on hydrogen rich syngas production from biomass gasification in a dual fluidized bed gasifier. Chemical Engineering and Processes. 2018;**126**

[73] Al-Farraji A, Marsch R, Steer J. A comparison of the pyrolysis of olive kernel biomass in fluidized and fixed bed conditions. Waste Biomass Valor. 2017;**8**:1273-1284

[74] Heidari A et al. Effect of process conditions on product yield and composition of fast pyrolysis of *Eucalyptus grandis* in fluidized bed reactor. 2014;**20**:2594-2602

[75] Papadikis K, Gu S, Bridgewater AV. CFD modelling of the fast pyrolysis of biomass in fluidized bed reactors: Modelling the impact of biomass shrinkage. Chemical Engineering Journal. 2009;**149**:417-427

[76] Zhu JY, Pan X, Zalesny RS. Pretreatment of woody biomass for biofuel production: Energy efficiency. Technologies, and Recalcitrance. 2010;**87**(3):847-857

[77] Himmel ME et al. Biomass recalcitrance: Engineering plants and enzymes for biofuels production. 2007;**315**(5813):804-807

[78] Mosier N et al. Features of promising technologies for pretreatment of lignocellulosic biomass. 2005;**96**(6):673-686

Chapter 5

Fungi for Bioeconomy: Possibilities and Perspectives

Maria Augusta Horta, Jaire Ferreira Filho and Anete Pereira de Souza

Abstract

Fungi represent one of Earth's most diverse and ecologically vital kingdoms, with an estimated 2.2–3.8 million species, yet only a small part formally described. Their remarkable metabolic versatility enables them to thrive in extreme environments, decompose complex organic compounds, and form symbiotic relationships critical for ecosystem functioning. This chapter explores fungal diversity through the lens of fungi possibilities, highlighting key species and the genetic mechanisms that confer fungi interesting characteristics. We examine fungal adaptability mechanisms—from heat-shock proteins in thermophiles to ligninolytic enzymes in white-rots—and their biotechnological applications in bioremediation, biofuels, and biopharmacos. Nowadays, the bioeconomy based on fungi products and subproducts evolved into cutting-edge engineered bioplatforms for sustainable bioprocesses and carbon-negative technologies. Further, we discuss how advances in genomic data analysis through cloud-based next-generation sequencing pipelines, linked to the growing computational processing capacity, are bringing even more possibilities to fungi discoveries and accelerating the transition to fungal-based technologies. Despite all potential, climate change and biodiversity loss threaten undiscovered species with potential industrial value and pose risks to agriculture and health. Thus, integrating multi-omics, synthetic biology, and conservation strategies will be crucial to harnessing fungi's full potential in addressing global challenges.

Keywords: fungi, fungus, diversity, bioeconomy, bioprocess, next generation sequencing, bioplatform

1. Introduction

This chapter aims to:

- Discuss fungal diversity and investigate how molecular phylogenetics, DNA studies, and high-throughput data analysis are improving the number of described and estimated species and revealing the genetic arsenal of the fungi kingdom.

- Decipher molecular adaptation mechanisms (extremophile biology, metabolic versatility, and symbiosis) that enable ecological success and biotechnological utility.

IntechOpen

- Assess fungi's historical, current, and future roles in the bioeconomy, showing recent studies that describe relevant bioproduction from fungi (bioethanol, biomolecules, enzymes, and others).
- Identify critical demands in fungal conservation, strain optimization, and regulatory frameworks for scaling bioproduction from fungi.

Key research questions:
Taxonomic and ecological dimensions:

- How do molecular tools redefine fungal biodiversity and cryptic species discovery?
- What genomic traits underpin niche specialization in different environments?

Biotechnological approaches:

- How do genetic data analysis and molecular-based engineering bridge lab-to-industry yield gaps?
- What are the fungi's derived bioproducts that present economic viability, and what are the promising candidates for new bioproducts from fungi?
- How has biotechnological production from fungi increased over history, and what are the perspectives for the future?

Sustainability challenges:

- How can fungi-derived bioprocess address carbon-negative applications for sustainable challenges?
- How do climate-induced range shifts alter fungal pathogen/benefactor balances in human health and agriculture?

Fungi represent one of the most diverse and ecologically vital kingdoms of life, with over 2.2 to 3.8 million species estimated globally [1, 2]. They live as thread-like structures or in single-celled form within the soil or other organic materials, in water or on rock, with a small percentage of species on occasion flaunting spore-making mushrooms, brackets, or other structures (**Figure 1**). Most fungi disperse their seeds, the spores, by air, animals, or water droplets. This requires the formation of spore-bearing structures, such as the familiar mushrooms, truffles, cup, and bracket fungi, molds, or rusts, among other structures proper to spread spores in the environment. Many of these structures are ephemeral and, in some cases, show seasonality.

In contrast, lichen-forming fungi have a perennial appearance, with permanent spore-bearing structures, as they feed on carbohydrates photosynthesized by their symbiotic partners (cyanobacteria and/or algae) and can withstand extreme desiccation [4, 5]. Fungi come in a vast array of sizes; they hold the record for being the largest organism on Earth (e.g., *Armillaria ostoyae*, spanning more than 10 km^2 in Oregon, USA [6]), but they can also have diminutive bodies hard to examine with

Figure 1.
Diversity of different types of fungi. (A) Phragmidium *sp. [rose rust], (B)* Calocera *sp., (C)* Trametes *sp., (D)* Tilletia *sp. [smut], (E)* Colletotrichum *sp. [Leaf spot], (F)* Erysiphe *sp. [powdery mildew cleistothecia], (G)* Inonotus *sp., (H)* Termitomyces *sp., (I)* Kweilingia *sp. [rust], (J)* Podosphaera *sp. on* Sonchus *sp. [powdery mildew], (K)* Tremella *sp., (L) Xylaria sp., (M) Uromyces sp. [aecia and telia], (N)* Pileolaria *sp. [rust], (O)* Gaestrum *sp., (P)* Didymium *sp., (Q)* Penicillium *sp. on* Emblica *sp., (R)* Schiffnerula *sp. [black mildew], (S)* Aspergillus *sp., (T)* Coleosporium *sp. [rust], (U)* Schizophyllum *sp., (V)* Aspergillus *sp. [on cow pea], (W)* Mitteriella *sp. [black mildew] and (X)* Periconia *sp. Scale bars A–X = 20 m [3].*

conventional morphometric techniques. Approximately 155,000 to 157,000 fungal species have been formally described. It represents only a small part of the estimated global fungal diversity [1, 5, 7]. Thus, fungi bring enormous potential to the discovery of new compounds. They play critical roles in ecosystems as decomposers, symbionts, and pathogens, driving nutrient cycling and forming mutualistic relationships with plants and algae.

Fungi diverged from animals over 1.3 billion years ago, evolving an arsenal of adaptations that allow them to thrive in every habitat on Earth, from deep-sea vents to arid deserts. Their remarkable metabolic versatility enables them to break down complex organic compounds, including lignin and cellulose, making them indispensable in carbon cycling. Additionally, fungi exhibit extreme tolerance to harsh conditions, such as high radiation, drought, and toxic metals, showcasing their evolutionary resilience and successful adaptative capacity [1].

Nowadays, climate change and habitat destruction threaten fungal biodiversity; while emerging fungal pathogens pose risks to agriculture and health. Understanding fungal metabolism, diversity, and adaptability is essential for biotechnology, conservation, and medicine, and further research into this enigmatic kingdom is necessary to improve fungi-based technologies.

1.1 Fungal and its role in bioeconomy: Past, present, and future perspectives

Fungi have played a crucial but often underappreciated role in the bioeconomy throughout human history. In the past, traditional societies relied on fungi for food (e.g., baking, brewing fermented beverages like beer and soy products, as the use of *Aspergillus oryzae* in sake production) and medicine (e.g., *Penicillium notatum* molds used in ancient wound treatments), though their scientific basis was unknown [8].

The twentieth century marked a turning point with the discovery of penicillin [9] and the industrial-scale production of fungal enzymes (primarily amylases for textiles, and cellulases) and pharmaceuticals. Today, solvents, antibiotics, enzymes, vitamins, amino acids, polymers, and many other useful compounds are produced on a large scale [10, 11]. Fungi are central to the circular bioeconomy, enabling the development of sustainable materials, such as mycelium-based leather [12], biofuels as the second-generation bioethanol [13], and bioremediation strategies, as the use of *Pleurotus* spp. degrading petroleum waste [14] and Fusarium degrading microplastics [15].

And that is not all. Cutting-edge advances in CRISPR-based fungal and synthetic biology are expanding applications, and the metabolic engineering techniques promise to create optimized cell factories, for different proposes [16]. Prospects include carbon-negative construction materials from mycelium and agricultural biofertilizers through improving mycorrhizal networks. However, challenges in strain scalability and optimization and biosafety regulations remain.

The fungal kingdom represents an extraordinary diversity of organisms with unparalleled metabolic diversity and profound impacts on animal, plant, and ecosystem health. Fungi present us with an arsenal of possibilities and are poised to drive the transition to a post-petroleum economy, acting in different scenarios, from bioproduct production to climate change mitigation [17–19].

2. Fungal diversity

Fungi exhibit extraordinary interspecies diversity reflected in the phylogenetic distribution (**Figure 2**). Major phyla like Ascomycota and Basidiomycota dominate terrestrial ecosystems, while Chytridiomycota and Cryptomycota reveal hidden diversity in aquatic environments [18]. This variation is driven by ecological specialization, with fungi occupying niches from Arctic permafrost to acidic mines. Thus, extremophilic fungi have been isolated worldwide from saltern brines, magnesium-rich bitterns, from the hypersaline waters of the Dead Sea, the Great Salt Lake, lakes in Antarctica, and the alkaline Wadi El Natrun in Egypt [21]. Even at hypersaline industrial effluents in temperate and tropical climates is possible to find fungi; the diversity observed across nature reflects a range of interspecies differences, evident in morphology (yeasts vs. hyphae), reproductive strategies, and metabolic capabilities [3, 20, 21].

At the intraspecies level, genetic and phenotypic plasticity enables rapid adaptation [22]. For example, *Cryptococcus neoformans* show strain-specific virulence due to genomic rearrangements [23], while *Aspergillus fumigatus* populations develop antifungal resistance via mutations [24]. Pathogenic fungi like *Candida auris* demonstrate alarming intraspecies diversity, with multidrug-resistant strains emerging independently across continents [24].

Such diversity arises from genetic occurrences such as horizontal gene transfer (e.g., toxin genes in *Fusarium* [25]), hybridization (yeast hybrids in brewing [26]), and transposable elements action. This genomic plasticity underpins fungal roles in

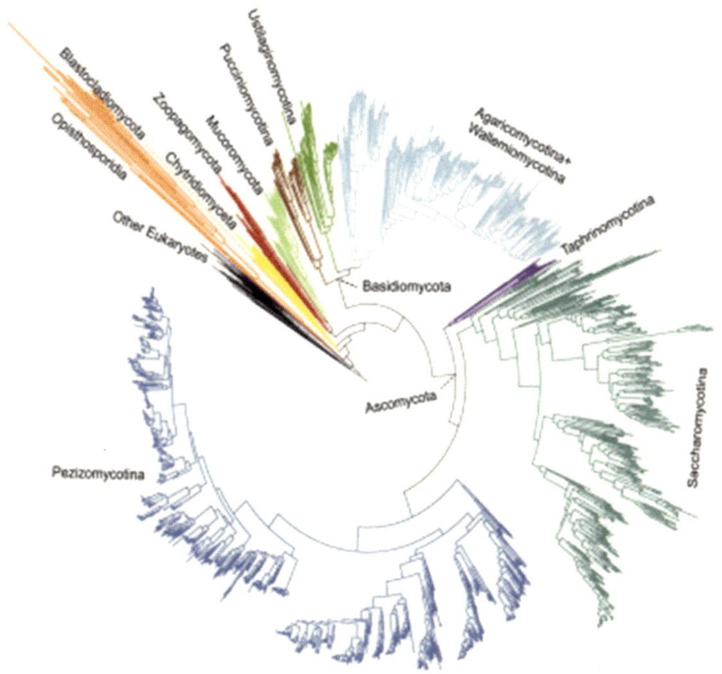

Figure 2.
Phylogenomic tree from a data matrix of 290 genes from genomes of 1644 fungi species that include representatives from most major fungal lineages [20].

biotechnology, pathogenesis, and ecosystem resilience. Nowadays genomics-driven classification improves the understanding of the genetic traits that influence fungi metabolism and adaptation.

The symbiosis between microbes and multicellular organisms is ubiquitous in life. The resulting changes in biosynthetic goods and services are recognized as sources of evolutionary innovation that allow systems to coevolutionary adaptations, and as a result, survive. Fungi have many symbiotic relationships with other organisms, including insects, plants, and algae. These relations act as a signaling language for environmental adaptation and surveillance, and thus, the major symbiotic relations aim to provide nutrients for both organisms. For example, *Rhizophagus irregularis* from mycorrhizal fungi population on soil can show host-specific genetic tuning to soil adaptation [27], endophytic fungi produce alkaloids protecting grasses from herbivores [28], and lichens exhibit holobiont diversity, where fungal partners select algal genotypes [29]. These examples underscore how selection pressures shape fungal diversity at micro- and macro-evolutionary scales.

Fungal diversity is a cornerstone of the emerging bioeconomy, offering unparalleled opportunities for sustainable innovation across industries. Recent studies highlight how fungi's metabolic versatility enables breakthroughs in bioremediation (e.g., fungi degrading plastics [15]), biofuel production (engineered *Trichoderma* strains for lignocellulosic ethanol) [13], and alternative protein sources (mycelium-based foods with high nutritional value [30]). Fungi also drive circular economy solutions, such as mycelium-based packaging materials, agricultural biofertilizers (mycorrhizal fungi enhancing crop resilience [31]), and pharmaceuticals (novel antifungals from understudied species) underscores their economic and ecological value. However, harnessing this potential requires preserving fungal biodiversity, as climate change threatens undiscovered species with industrial applications. Advances in metagenomics and synthetic biology are accelerating fungal bioprospecting, positioning fungi as indispensable to a post-fossil-fuel future [32].

Why this matters

- Sustainability: Fungi can replace petroleum-based materials (plastics, fuels).
- Food security: Mycelium proteins can address global protein demand.
- Medical innovation: Underexplored species can yield new antibiotics.
- Climate resilience: Mycorrhizal fungi can sequester carbon and enhance soil health.

2.1 Major phyla molecular phylogeny

Fungi are classified into several major phyla based on molecular phylogenetics (**Figure 2**), with Ascomycota and Basidiomycota dominating terrestrial ecosystems. Ascomycota includes over 64,000 described species, such as Penicillium and Morchella, and is renowmed, for its roles in decomposition, plant pathogenesis, and antibiotic production [33]. Basidiomycota, with 30,000+ species, encompasses mushrooms, rusts, and mycorrhizal fungi, playing critical roles in nutrient cycling and symbiotic plant interactions [34]. Early-diverging lineages like Chytridiomycota (aquatic, flagellated fungi) and Zygomycota (e.g., Rhizopus) are now recognized as polyphyletic, requiring reclassification via genomic data [35] while Glomeromycota, species-poor (~300 spp.), are ecologically vital as arbuscular mycorrhizal symbionts [36].

Fungal diversity is vastly underestimated due to cryptic species—morphologically identical but genetically distinct lineages. Traditional identification based on spores or hyphal structures often fails to distinguish species, whereas DNA barcoding (e.g., ITS region) reveals hidden diversity in the genome, and high-throughput molecular identification techniques have greatly progressed the understanding of the diversity of mutualists, saprotrophs, and pathogens fungi. Environmental DNA (eDNA) studies highlight this disparity. In soil fungi, 40% of operational taxonomic units (OTUs) in tropical soils lack matches to described species [37]. The advances due to molecular techniques reach, for example, Lichens identification of single morphospecies, like *Sticta fuliginosa* presenting more than 20 genetic lineages described [38], and in Marine fungi, metabarcoding approaches uncover novel chytrids in ocean microbiomes [39]. These findings underscore the potential of integrative taxonomy, combining genomics, ecology, and morphology. This approach can reveal cryptic species with unique ecological roles, assist biodiversity conservation, reveal undescribed fungi proper for novel enzymes or drug applications in biotechnological platforms, and accurately improve species delimitation, necessary for ecosystem management.

2.2 Mechanisms of fungal adaptability: From molecular strategies to ecological success

Fungi exhibit remarkable adaptability, allowing them to thrive in diverse and extreme environments. Thus, fungi developed different strategies to adapt and survive. Fungi can be grouped according to adaptation strategy on "extreme environments, metabolic versatility, and symbiotic adaptations," each one using a specific key cellular mechanism, with molecular basis described by different works. **Table 1**

Adaptation type	Example organisms	Key mechanisms	Cellular adaptation basis	References
Extreme environments				
Thermophiles	*S. cerevisiae*	Heat-shock proteins (HSP70/90), melanin production	Protein stabilization, ROS scavenging	[40]
Psychrophiles	*Pseudogymnoascus* spp.	Antifreeze proteins, unsaturated fatty acids	Ice crystal inhibition, membrane fluidity maintenance	[41, 42]
Xerophiles	*Xeromyces bisporus*	Trehalose accumulation, hyphal melanization	Osmoprotection, UV/ desiccation resistance	[43, 44]
Metabolic versatility				
Lignin degradation	*Phanerochaete chrysosporium*	Laccases, peroxidases (LiP/MnP), Fenton chemistry	Aromatic ring cleavage, hydroxyl radical generation	[45, 46]
Heavy metal tolerance	*Fusarium oxysporum*	Metallothioneins, ABC transporters	Metal chelation, vacuolar sequestration	[47, 48]
Symbiotic adaptations				
Mycorrhizae	*Rhizophagus irregularis*	SYM pathway, enzyme secretion (phosphatases)	Nutrient exchange (P/N), host communication	[49, 50]
Lichens	*Clodonia rangifering*	Parietin production, haustorial interfaces	UV shielding, photosynthate transport	[5, 51]

Table 1.
Adaptation strategies used by fungi, grouped by classes according to different mechanisms of cellular adaptation.

shows the adaptation strategies and the corresponding examples of organisms that developed key cellular mechanisms based on molecular adaptations.

In extreme habitats, thermophiles can produce heat-shock proteins (HSP70/90) that maintain protein stability, while psychrophiles such as *Pseudogymnoascus* spp. survives in subzero temperatures by synthesizing antifreeze proteins to prevent ice crystal formation. Xerophiles, including desert crust fungi, accumulate trehalose and melanin to resist desiccation and UV radiation or produce protective melanin [43]. Metabolically, fungi degrade complex substrates, like lignin by using such as manganese peroxidase (MnP), lignin peroxidase (LiP), and versatile peroxidase (VP) [13], while heavy metal-tolerant species employ chelation and vacuolar sequestration to detoxify pollutants [45, 47]. Symbiotic fungi, such as mycorrhizae, enhance nutrient exchange with plants via specialized signaling pathways [49], and lichen-forming fungi protect photosynthetic partners through UV-shielding pigments [51]. These adaptive strategies underscore fungi's ecological versatility and highlight their potential in bioremediation, agriculture, and biotechnology, making them indispensable to ecosystem resilience and sustainable innovation.

Nowadays, cutting-edge omics technologies and synthetic biology tools precisely reveal the remarkable adaptive strategies specially of extremophilic fungi, creating unprecedented opportunities for biotechnology. While new extremophile species are described, the integration of multi-omics data analysis with machine learning has accelerated the understanding of the metabolic versatility of extremophiles, leading to the discovery of enzymes, exopolysaccharides, and various types of compounds that find applications in diverse industries [52, 53]. Practical applications should emerge, solutions for sustainable agriculture, high-salinity bioremediation, environmental cleanup, and biomanufacturing in extreme conditions could benefit from the deep understanding of extremophile unique mechanisms.

3. Human impacts and applications of fungi: From threats to solutions

Due to the vital ecology and the great diversity, the fungal kingdom profoundly influence human activities, presenting both challenges and opportunities. As agricultural pathogens like *Fusarium graminearum* devastate crops through mycotoxin contamination, they threaten global food security. Conversely, fungi are indispensable to industry, producing life-saving antibiotics (e.g., penicillin) and enzymes for food processing and biofuels [10, 13]. Climate change is altering fungal distributions, expanding the ranges of pathogenic species while disrupting beneficial symbioses. Yet, fungi can also offer hundreds of sustainable solutions. We can cite the mycoremediation cleaning for polluted environments, and mycelium-based materials replacement plastics [12, 14]. Thus, the dual roles of fungi as threats and allies underscore the need for balanced management—harnessing their potential while mitigating risks in a changing world.

3.1 Fungi as powerful microbial platforms for sustainable bioproduction

Fungi have emerged as indispensable microbial platforms for producing biofuels, biochemicals, pharmaceuticals, and biomaterials, owing to their metabolic versatility, genetic tractability, and ability to utilize diverse feedstocks. A prime example of success is the yeast *Saccharomyces cerevisiae*, which dominates global bioethanol production, generating over 130 billion liters annually—accounting for ~90% of the world's biofuel output, with an estimated market size of $100 billion dollars [54]. This platform has been enhanced through genetic engineering, including the introduction of xylose isomerase pathways to ferment lignocellulosic biomass, boosting yields by 20–30% [55]. Another breakthrough is *Pichia pastoris*, a workhorse for recombinant protein production, manufacturing insulin, vaccines, and monoclonal antibodies [56, 57]. Filamentous fungi like *Aspergillus niger* contribute significantly to the organic acid market, producing 2.2 million tons of citric acid annually—a key ingredient in food, pharmaceuticals, and biodegradable plastics [58]. Emerging platforms like *Yarrowia lipolytica* (for lipid-based biofuels) and *Trichoderma reesei* (for low-cost enzymes) highlight fungi's expanding potential to replace petroleum-dependent processes with sustainable, high-yield alternatives. **Table 2** estimates the actual market size of main derived fungi products, with examples in pharmaceutical, biofuel, food and agriculture, biomaterials, and bioremediation sectors.

Key bioproducts from fungi:

- *Pichia pastoris*: workhorse for therapeutic proteins.
- *Saccharomyces cerevisiae*: bioethanol top producer.
- *Aspergillus niger*: citric acid main producer.
- *Trichoderma reesei*: enzymes producer ongoing on cost reduction with edited strains.

Market sector	Key fungal products	Estimate global market size (USD/year 2023–2024)	Notable strains/ applications	References
Pharmaceuticals	Antibiotics (penicillin, statins)	$70 billion	*Penicillium chrysogenum, Asporgittus terreus*	[58]
	Recombinant proteins (insulin, vaccines)	$20+ billion	*Pichia pastoris*	[57]
Industrial enzymes	Cellulases. amylases, lipases	$10 billion	*Thichoderma reesei, Aspergillus niger*	[59]
Biofuels	Bioethanol	$100 billion	*Saccharomyces cerevisiae*	[54]
	Biodiesel (microbial lipids)	$4.2 billion	*Yarrowia lipolytica*	[60]
Food and agriculture	Mycelium-based meat (Quorn)	$1.3 billion	*Fusarium venenatum*	[61]
	Mycorrhizal biofertilizers	$2.8 billion	*Rhizophagus irregularis*	[62]
Biomaterials	Mycelium packaging/ leather	$300 million	*Ganoderma, Trametes*	[12]
Bioremediation	Hydrocarbon degradation	$1.5 billion	*Pleurotus ostreatus*	[63]

Table 2.
Estimates of the market size of fungal products in variated sectors (USD/year 2023–2024).

Fungi are prolific producers of enzymes that drive industrial bioprocesses, enabling the sustainable production of valuable products. Species like *Aspergillus* and *Trichoderma* secrete hydrolytic enzymes (e.g., cellulases, amylases, and proteases) that break down complex biomass into fermentable sugars for biofuel production [13]. White-rot fungi, such as *Phanerochaete chrysosporium*, produce ligninolytic enzymes (laccases, peroxidases) that degrade plant biomass, facilitating waste-to-resource conversion in biorefineries [45, 46]. These enzymes operate under mild conditions, reducing energy consumption compared to chemical methods, and are increasingly optimized through genetic engineering and fermentation technology [32, 64].

Beyond biofuels, fungal enzymes are pivotal in the food, pharmaceutical, and textile industries. For example, *Rhizopus* spp. generates lipases used in dairy flavor enhancement, while *Penicillium* spp. synthesizes glucose oxidase for biosensors and food preservation. Fungal pectinases and tannases improve juice clarification and wine processing, demonstrating their versatility in the food industry [10].

Nowadays, fungi as a potential source of enzymes is an acceptable idea. The next step to maximize enzyme yield production seems to be on the development of fermentative techniques, such as solid-state and submerged fermentation, with advances in bioreactor design and strain engineering further enhancing efficiency. The scalability of these bioprocesses underscores their potential to replace petroleum-dependent production systems [17].

Emerging applications include fungal-derived biodegradable plastics (e.g., polyhydroxyalkanoates from *Aspergillus*), biosurfactants for oil spill remediation, and enzyme cascades for fine chemical synthesis. CRISPR-based genome editing now tailors fungal strains for higher enzyme titers and novel functionalities [65, 66].

As circular bioeconomy principles gain traction, bioprocesses that use bioreactions, such as those performed by fungi, offer a roadmap to decarbonize industries—turning agricultural residues, wastewater, and CO_2 into high-value products while minimizing environmental impact [67].

3.2 Advances in bioprocess scaling and optimization for fungal-based production

Recent studies have demonstrated significant progress in scaling fungal bioprocesses from lab to industrial levels through optimized culture conditions and media design. However, scaling up bioreactor systems for fungal fermentation poses significant challenges, including oxygen transfer limitations, shear stress sensitivity, and heterogeneous nutrient distribution, particularly due to the filamentous morphology [68]. High broth viscosity in fungal cultures can difficult mixing and mass transfer, and to address these issues, segmented airflow systems and modified impeller geometries have been implemented to enhance oxygen dispersion while minimizing hyphal damage [69]. New solutions are been proposed, as 3D-printed static mixers in stirred-tank bioreactors, which enhance oxygen transfer while reducing energy consumption during fermentation [70]. For shear-sensitive fungi (e.g., *Trichoderma reesei*), novel bioreactor designs with segmented airflow and adaptive impellers mitigate hyphal damage during scale-up, improving oxygen transfer efficiency [71]. Additionally, advanced process control strategies, including fed-batch cultivation with real-time monitoring of dissolved oxygen and pH, have been crucial in scaling up. The recent integration of machine learning (ML)-driven adaptive control has improved real-time monitoring, optimizing substrate feeding and reducing product accumulation [72]. The fine control of the bioprocess now enables rapid identification of ideal growth parameters (pH, temperature, and dissolved O_2) for filamentous fungi, maximizing enzyme or metabolite yields while minimizing energy inputs [73]. For example, the modular development of media formulations—incorporating agro-industrial waste (e.g., rice bran and sugarcane bagasse) as low-cost substrates—has reduced production costs by 30–40% while maintaining high titers [73].

These innovations highlight how integrating bioprocess engineering with fungal physiology can overcome key scalability challenges in industrial biotechnology.

3.3 Cutting-edge genetic modifications in fungal bioproduction platforms

Recent advances in genetic engineering have revolutionized fungal biotechnology allowing precise modifications on the genetic machinery to improve enzyme production, metabolic pathways, and stress resistance. Homologous recombination has been widely used to integrate heterologous pathways, such as introducing xylose metabolism genes into *Saccharomyces cerevisiae* for improved bioethanol yield [74]. Additionally, gene editing via base editors has optimized fungal strains without introducing double-strand breaks—*Aspergillus niger* was modified to overexpress citric acid biosynthetic genes [75]. Stress resistance has also been enhanced through genetic modifications; for instance, *S. cerevisiae* strains with engineered heat-shock proteins (HSPs) exhibit improved thermotolerance in industrial fermentations [76]. These techniques collectively expand fungal applications in biotechnology, enabling sustainable production of enzymes, biofuels, and pharmaceuticals.

More recently CRISPR-Cas9 has emerged as a powerful tool, allowing targeted gene knockouts, insertions, and base editing in fungi. In *Saccharomyces cerevisiae*, researchers achieved a breakthrough by performing multiplexed CRISPR edits to simultaneously knockout genes, significantly enhancing ethanol tolerance and fermentation efficiency [77]. Similarly, *Aspergillus niger* has been engineered using novel base-editing systems to create comprehensive promoter libraries [75]. Similarly, *Neurospora crassa* shows great promise through optimized CRISPR-Cas9 editing, for significantly improving production characteristics [78].

Beyond traditional editing tools, sophisticated biosensing systems are being implemented, as well as quorum-sensing circuits that precisely synchronize production with biomass feeding. These cutting-edge modifications demonstrate how precision genetic tools are pushing the boundaries of fungal bioproduction capabilities. These applications highlight fungi's versatility, with CRISPR, promoter engineering, and pathway refactoring driving innovations in biomanufacturing, medicine, and sustainable chemistry.

4. High-potential fungi and future frontiers in fungal biotechnology

The field of fungal biotechnology is witnessing remarkable progress with emerging species demonstrating exceptional bioproduction capabilities. *Yarrowia lipolytica* has emerged as a particularly versatile platform, with engineered strains for biodiesel production [79], while other variants achieve impressive fatty acid yields exceeding 90 g/L through valorization of cooking oil waste [80]. The oleaginous yeast *Rhodotorula toruloides* demonstrates dual functionality within efficient sugar utilization on the bioproduction of lipids from lignocellulosic substrates, for microbial oil production [81].

Looking ahead, the integration of artificial intelligence with fungal biotechnology is opening new frontiers, including the use of generative AI to design specific proteins, unlocking opportunities in therapeutics, diagnostics, and synthetic biology. Real-time monitoring sensors and AI-driven feedback further optimize bioprocesses in different stages [53, 82].

These advancements, coupled with CRISPR-engineered hyper-producing strains, are bridging the lab-to-factory gap, with recent pilot plants achieving unprecedented higher production titers. The innovation following AI in the bioprocess field is only beginning. Unimaginable possibilities are in progress, as the recent observation of researchers that microelectronic devices could directly communicate with biology, as electronic information can be transmitted via redox reactions within biological systems [73].

4.1 Fungi in the next-generation sequencing (NGS) era

The advent of high-throughput DNA sequencing technologies has revolutionized fungal biotechnology by enabling rapid, cost-effective genome analysis, strain optimization, and functional genomics. Fungi play a pivotal role in this NGS-driven era, as their genomes provide critical insights into enzyme production, secondary metabolite biosynthesis, and stress adaptation. Whole-genome sequencing (WGS) of industrially relevant fungi, such as *Trichoderma reesei* and *Aspergillus niger*, has identified key cellulase and organic acid biosynthesis genes, facilitating targeted CRISPR-Cas9 editing for enhanced yields [83]. RNA-seq transcriptomics has elucidated regulatory networks and metagenomic sequencing of fungal communities in extreme environments (e.g., acidic mines, Arctic soils) and has uncovered novel ligninolytic enzymes and cold-adapted proteases with industrial applications [84, 85]. NGS also aids in pathogen surveillance, detecting antifungal resistance mutations in impactful fungi for human health, such as *Candida auris* [86] and *Aspergillus fumigatus* [87].

By integrating multi-omics (genomics, transcriptomics, and proteomics) analysis, researchers are engineering fungal strains for sustainable bioproduction, from waste to biofuels to precision fermentation of drugs. The NGS age has thus transformed fungi into genomically empowered cell factories, driving innovations in the bioeconomy.

NGS also plays a pivotal role in biodiversity studies, with metagenomic sequencing of soil and marine microbiomes uncovering unculturable fungal species with unique enzymatic capabilities [2, 88]. By integrating transcriptomics (RNA-seq) and epigenomics (ChIP-seq), researchers dissect regulatory networks controlling enzyme production, such as the xyr1 transcription factor in *Trichoderma*, which modulates cellulase expression [89].

These advances underscore NGS as a cornerstone of fungal biotechnology, enabling precision strain development and ecological conservation.

4.2 Computational methods in fungal NGS technologies: From genome assembly to functional prediction

The analysis of fungal next-generation sequencing (NGS) data relies on advanced computational pipelines to assemble, annotate, and interpret complex genomic information (**Figure 3**). De novo genome assembly of fungi is typically performed using tools like SPAdes or Canu, which handle repetitive regions and high heterozygosity common in filamentous fungi [90]. For annotation, BRAKER and Funannotate leverage machine learning to predict genes, non-coding RNAs, and regulatory elements, integrating RNA-seq data for improved accuracy [91]. Comparative genomics platforms like OrthoFinder identify conserved and species-specific genes, revealing horizontal gene transfer events in Aspergillus and *Trichoderma* that enhance lignocellulose degradation [92]. Variant calling (e.g., using GATK or BCFtools) detects SNPs and indels linked to traits like enzyme production or antifungal resistance [93]. For metagenomic studies, tools such as MetaPhlAn and Kraken2 classify fungal taxa in environmental samples, while antiSMASH predicts secondary metabolite clusters [94]. Concurrently, fungi data is processed, and machine learning models are created to assist in the understanding of fungi interactions, protein secretion signals, and other characteristics that can guide synthetic biology [95, 96]. It is also transforming fungal genome annotation.

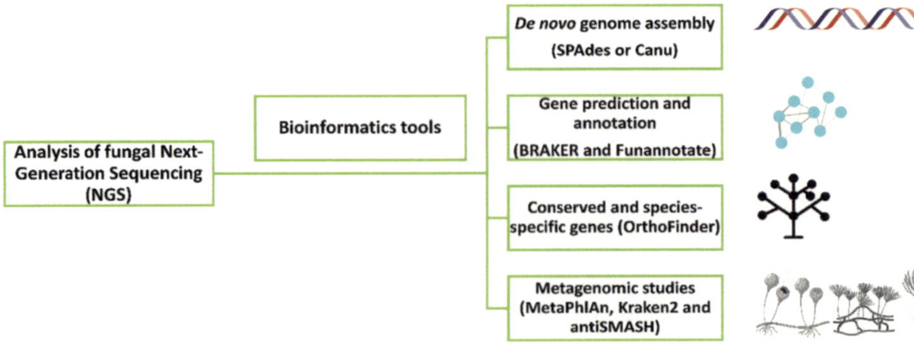

Figure 3.
Bioinformatics tools used for various analyses with next-generation sequencing data.

Tools like Funannotate-Train use convolutional neural networks (CNNs) to predict promoter regions and secretory signals with >90% accuracy [91, 97].

The deep analysis of NGS has revolutionized fungal genomics, especially strain optimization by enabling precision genome editing, informed by real-time genomic data. These computational advances accelerate the translation of fungal NGS data into biotechnological applications, from engineered strains to ecological insights.

4.3 Cloud-based NGS analysis and pangenome approaches for fungi: Scalable solutions for genomic insights

The exponential growth of fungal sequencing data has driven the adoption of cloud-based NGS pipelines, enabling scalable, collaborative, and cost-efficient genome analysis. Platforms like Google Cloud Life Sciences and AWS Genomics CLI provide optimized workflows for genome assembly, RNA-seq analysis, and metagenomics. The high processing capacity allowed the emergence of fungal pangenomics, a comparative analysis of core (shared) and accessory (individual or present at groups) orthologs genes across strains that unlock evolutionary and functional trades. For example, pangenome analysis of *Aspergillus fumigatus* identifies genomic variation associated with clinical isolates and resistance as well as genetic variation in virulence factors [97]. For pangenome analysis, key tools are been used, such as Panaroo, which detects gene presence/absence variations (PAVs) in fungal populations, and Anvi'o, to visualize pangenomes and identify horizontal gene transfer [98].

The impact of the high throughput analyses can be observed in the biotechnological processes' rapid development, with consistent results applied to bioprocess even more sustainable.

5. Conclusion

Fungi's unparalleled genetic and metabolic diversity positions them as indispensable players in ecosystem resilience and the emerging bioeconomy. From their roles as ancient decomposers to modern cell factories producing biofuels, enzymes, and biopharmaceuticals, fungi demonstrate extraordinary adaptability through molecular innovations like extremophile proteins and symbiotic signaling networks. While genomic tools have uncovered cryptic diversity and enabled strain engineering, most fungal species remain undescribed, with their ecological and economic potential largely untapped. Climate change and habitat destruction necessitate urgent conservation efforts alongside bioprospecting initiatives. Future research must bridge gaps in fungal taxonomy, scale up AI-driven bioprocess optimization, and address biosafety regulations to fully realize fungi's capacity to drive sustainable innovation. As humanity transitions toward post-petroleum economies, fungi—with their 1.3 billion years of evolutionary ingenuity—offer solutions for carbon sequestration, circular materials, and climate-adaptive agriculture, underscoring the critical need to protect and study this enigmatic kingdom.

Acknowledgements

Authors thanks to University of Campinas (UNICAMP) and São Paulo Research Foundation (FAPESP) for the research support (FAPESP process 2014/18856-1).

Conflict of interest

The authors declare no conflict of interest.

Author details

Maria Augusta Horta[1*], Jaire Ferreira Filho[1] and Anete Pereira de Souza[2]

1 Center for Molecular Biology and Genetic Engineering (CBMEG), University of Campinas (UNICAMP), Campinas, Brazil

2 Department of Plant Biology, Biology Institute, University of Campinas (UNICAMP), Campinas, Brazil

*Address all correspondence to: mariaaugusta.horta@gmail.com

References

[1] Hawksworth DL, Lücking R. Fungal diversity revisited: 2.2 to 3.8 million species. Microbiology Spectrum. 2017;**5**(4):1-17

[2] Tedersoo L, Bahram M, Põlme S, Kõljalg U, Yorou NS, Wijesundera R, et al. Global diversity and geography of soil fungi. Science (80-). 2014;**346**(6213):1256688. Available from: https://www.science.org/doi/abs/10.1126/science.1256688

[3] Gautam AK, Verma RK, Avasthi S, Sushma, Bohra Y, Devadatha B, et al. Current insight into traditional and modern methods in fungal diversity estimates. Journal of Fungi. 2022;**2**:226. DOI: 10.3390/jof8030226

[4] Niskanen T, Lücking R, Dahlberg A, Gaya E, Suz LM, Mikryukov V, et al. Pushing the frontiers of biodiversity research: Unveiling the global diversity, distribution, and conservation of fungi. Annual Review of Environment and Resources. 2023;**48**:149-176. DOI: 10.1146/annurev-environ-112621-090937

[5] Spribille T, Resl P, Stanton DE, Tagirdzhanova G. Evolutionary biology of lichen symbioses. The New Phytologist. 2022;**234**(5):1566-1582

[6] Anderson JB, Bruhn JN, Kasimer D, Wang H, Rodrigue N, Smith ML, et al. Clonal evolution and genome stability in a 2500-year-old fungal individual. Proceedings of the Royal Society B. 2018;**285**:1-6

[7] Paul J, Gregory SÆ. An estimate of the lower limit of global fungal diversity. Biodiversity and Conservation. 2007;**16**:99-111

[8] Bennett JW. Mycotechnology: The role of fungi in biotechnology. Journal of Biotechnology. 1998;**66**(2):101-107 Based on a lecture held at the symposium, 'Progress in US Biotechnology', at the 8th European Congress on Biotechnology (ECB8) in Budapest, Hungary, August 1997.1. Available from: https://www.sciencedirect.com/science/article/pii/S0168165698001333

[9] Fleming A. On the antibacterial action of cultures of a penicillium, with special reference to their use in the isolation of *B. influenzæ*. British Journal of Experimental Pathology. 1929;**10**:226-236

[10] Adrio JL, Demain ÆAL. Fungal biotechnology. International Microbiology. 2003;**6**:191-199

[11] Flickinger MC, Drew SW, Creighton TE, Gaden E, Humphrey AE, Lydersen BK, et al. Encyclopedia of bioprocess technology: Fermentation, biocatalysis, and bioseparation. In: Wiley Biotechnology Encyclopedias. New York, N.Y.: John Wiley & Sons, Inc; 1999

[12] Jones M, Gandia A, John S, Bismarck A. Leather-like material biofabrication using fungi. Nature Sustainability. 2021;**4**(1):9-16. DOI: 10.1038/s41893-020-00606-1

[13] Jouzani GS, Tabatabaei M, Aghbashlo M. Fungi in Fuel Biotechnology. Switzerland: Springer; 2020

[14] Kulshreshtha S. Removal of pollutants using spent mushrooms substrates. Environmental Chemistry Letters. 2019;**17**(2):833-847. DOI: 10.1007/s10311-018-00840-2

[15] Ekanayaka AH, Tibpromma S, Dai D, Xu R. A review of the fungi that degrade plastic. Journal of Fungi. 2022;**8**:1-27

[16] Salazar-Cerezo S, de Vries RP, Garrigues S. Strategies for the development of industrial fungal producing strains. Journal of Fungi (Basel, Switzerland). 2023;**9**(8):834

[17] Case NT, Berman J, Blehert DS, Cramer RA, Cuomo C, Currie CR, et al. The future of fungi: threats and opportunities. G3: Genes, Genomes, Genetics. 2022;**12**(11):jkac224

[18] Ali I, Qaiser H, Abdullah R, Kaleem A, Iqtedar M, Iqbal I, et al. Prospective roles of extremophilic fungi in climate change mitigation strategies. Journal of Fungi. 2024;**10**:385

[19] Lange L. The importance of fungi and mycology for addressing major global challenges*. IMA Fungus. 2014;**5**(2):463-471

[20] Li Y, Steenwyk JL, Chang Y, Wang Y, James TY, Jason E, et al. A genome-scale phylogeny of the kingdom fungi yuanning. Current Biology. 2021;**31**(8):1653-1665

[21] Gostincar C, Gunde-cimerman N. Understanding fungi in glacial and hypersaline environments. Annual Review of Microbiology. 2023;**77**:89-109

[22] Janbon G, Quintin J, Lanternier F. Studying fungal pathogens of humans and fungal infections: Fungal diversity and diversity of approaches. Microbes and Infection. 2019;**21**(5-6):237-245. DOI: 10.1016/j.micinf.2019.06.011

[23] Rizzo J, Wong SSW, Gazi AD, Moyrand F, Chaze T, Commere P-H, et al. Cryptococcus extracellular vesicles properties and their use as vaccine platforms. Journal of Extracellular Vesicles. 2021;**10**(10):e12129. Available from: https://isevjournals.onlinelibrary.wiley.com/doi/abs/10.1002/jev2.12129

[24] Fisher MC, Alastruey-Izquierdo A, Berman J, Bicanic T, Bignell EM, Bowyer P, et al. Tackling the emerging threat of antifungal resistance to human health. Nature Reviews. Microbiology. 2022;**20**(9):557-571

[25] Zhang Y, Yang H, Turra D, Zhou S, Ayhan DH, DeIulio GA, et al. The genome of opportunistic fungal pathogen fusarium oxysporum carries a unique set of lineage-specific chromosomes. Communications Biology. 2020;**3**(1):50. DOI: 10.1038/s42003-020-0770-2

[26] Burini J, Eizaguirre JI, Loviso C, Libkind D. Non-conventional yeasts as tools for innovation and differentiation in brewing. Revista Argentina de Microbiología. 2021;**53**:359-377

[27] Tanaka S, Hashimoto K, Kobayashi Y, Yano K, Maeda T, Kameoka H, et al. Asymbiotic mass production of the arbuscular mycorrhizal fungus Rhizophagus clarus. Communications Biology. 2022;**5**(1):43. DOI: 10.1038/s42003-021-02967-5

[28] Schardl CL, Leuchtmann A, Spiering MJ. Symbioses of grasses with seedborne fungal endophytes. Annual Review of Plant Biology. 2004;**55**:315-340. Available from: https://www.annualreviews.org/content/journals/10.1146/annurev.arplant.55.031903.141735

[29] Smith HB, Dal Grande F, Muggia L, Keuler R, Divakar PK, Grewe F, et al. Metagenomic data reveal diverse fungal and algal communities associated with the lichen symbiosis. Symbiosis. 2020;**82**(1):133-147. DOI: 10.1007/s13199-020-00699-4

[30] Wang X, Pei Y, Wu J, Zhong X, Liu H, Xue Y. Innovative mycelium-based food: Advancing One Health through

nutritional insights and environmental sustainability. Comprehensive Reviews in Food Science and Food Safety. 2025;**24**(2):e70166. Available from: https://ift.onlinelibrary.wiley.com/doi/abs/10.1111/1541-4337.70166

[31] Zhao G, Zhu X, Zheng G, Meng G, Dong Z, Baek JH, et al. Development of biofertilizers for sustainable agriculture over four decades (1980-2022). Geography and Sustainability. 2024;**5**(1):19-28. Available from: https://www.sciencedirect.com/science/article/pii/S2666683923000652

[32] Saranya S, Thamanna L, Chellapandi P. Unveiling the potential of systems biology in biotechnology and biomedical research. Systems Microbiology and Biomanufacturing. 2024;**4**(4):1217-1238. DOI: 10.1007/s43393-024-00286-4

[33] Spatafora JW, Aime MC, Grigoriev IV, Martin F, Stajich JE, Blackwell M. The fungal tree of life: From molecular systematics to genome-scale phylogenies. Microbiology Spectrum. 2017;**5**(5):1-34

[34] Naranjo-Ortiz MA, Gabaldón T. Fungal evolution: Cellular, genomic and metabolic complexity. Biological Reviews of the Cambridge Philosophical Society. 2020;**95**(5):1198-1232

[35] Li Y, Steenwyk JL, Chang Y, Wang Y, James TY, Stajich JE, et al. A genome-scale phylogeny of the kingdom fungi. Current Biology. 2021;**31**(8):1653-1665.e5. DOI: 10.1016/j.cub.2021.01.074

[36] van der Heijden MGA, Martin FM, Selosse M-A, Sanders IR. Mycorrhizal ecology and evolution: The past, the present, and the future. The New Phytologist. 2015;**205**(4):1406-1423

[37] Tedersoo L, Bahram M, Põlme S, Kõljalg U, Yorou NS, Wijesundera R, et al. Fungal biogeography. Global diversity and geography of soil fungi. Science. 2014;**346**(6213):1256688

[38] Magain N, Tniong C, Goward T, Niu D, Goffinet B, Sérusiaux E, et al. Species delimitation at a global scale reveals high species richness with complex biogeography and patterns of symbiont association in Peltigera section Peltigera (lichenized Ascomycota: Lecanoromycetes). Taxon. 2018;**67**(5):836-870. Available from: https://onlinelibrary.wiley.com/doi/abs/10.12705/675.3

[39] Grossart H-P, Van den Wyngaert S, Kagami M, Wurzbacher C, Cunliffe M, Rojas-Jimenez K. Fungi in aquatic ecosystems. Nature Reviews. Microbiology. 2019;**17**(6):339-354. DOI: 10.1038/s41579-019-0175-8

[40] Tiwari S, Thakur R, Shankar J. Role of heat-shock proteins in cellular function and in the biology of fungi. Biotechnology Research International. 2015;**2015**(1):132635. DOI: 10.1155/2015/132635

[41] Childress MK, Dragone NB, Young BD, Adams BJ, Fierer N, Quandt CA. Three new *Pseudogymnoascus* species (*Pseudeurotiaceae*, *Thelebolales*) described from Antarctic soils. IMA Fungus. 2025;**16**:e142219. DOI: 10.3897/imafungus.16.142219

[42] Leushkin EV, Logacheva MD, Penin AA, Sutormin RA, Gerasimov ES, Kochkina GA, et al. Comparative genome analysis of *Pseudogymnoascus* spp. reveals primarily clonal evolution with small genome fragments exchanged between lineages. BMC Genomics. 2015;**16**(1):400

[43] Sterflinger K, Tesei D, Zakharova K. Fungi in hot and cold deserts with particular reference to microcolonial

fungi. Fungal Ecology. 2012;**5**(4):453-462. Available from: https://www.sciencedirect.com/science/article/pii/S1754504811001437

[44] Gostinčar C, Stajich JE, Gunde-Cimerman N. Extremophilic and extremotolerant fungi. Current Biology. 2023;**33**(14):R752-R756. DOI: 10.1016/j.cub.2023.06.011

[45] Janusz G, Skwarek E. Potential of laccase as a tool for biodegradation of wastewater micropollutants. Water. 2023;**15**:3770

[46] van der Made JJA, Landis EA, Deans GT, Lai RA, Chandran K. Synergistic lignin degradation between *Phanerochaete chrysosporium* and Fenton chemistry is mediated through iron cycling and ligninolytic enzyme induction. Science of the Total Environment. 2023;**905**:166767. Available from: https://www.sciencedirect.com/science/article/pii/S0048969723053925

[47] Yao H, Shi W, Wang X, Li J, Chen M, Li J, et al. The root-associated fusarium isolated based on fungal community analysis improves phytoremediation efficiency of *Ricinus communis* L. in multi metal-contaminated soils. Chemosphere. 2023;**324**:138377. Available from: https://www.sciencedirect.com/science/article/pii/S0045653523006446

[48] Sharma A, Sharma P, Deepak B, Mona S. Microbes as environmental allies: From wastewater treatment to sustainable biofuel production. In: Bala K, Ghosh T, Kumar V, Sangwan P, editors. Harnessing Microbial Potential for Multifarious Applications. Singapore: Springer Nature Singapore; 2024. pp. 359-376. DOI: 10.1007/978-981-97-1152-9_14

[49] Aparicio V, Van DJ, Goormachtig S. Characterization of arbuscular mycorrhizal effector proteins. International Journal of Molecular Sciences. 2023;**24**:9125

[50] Phytologist N, Issue S. Editorial mycorrhizal research now: From the micro- to the macro-scale. The New Phytologist. 2024;**242**:1399-1403

[51] Pichler G, Grube M, Muggia L, Carniel FC, Kranner I. How to build a lichen: From metabolite release to symbiotic interplay. The New Phytologist. 2023;**238**:1362-1378

[52] Saini P, Mishra P. Bioprospecting of extremophiles for industrial enzymes. In: Saini P, Mishra P, editors. Bioprospecting of Microbial Resources for Agriculture, Environment and Bio-Chemical Industry. Cham: Springer Nature Switzerland; 2024. pp. 43-74. DOI: 10.1007/978-3-031-63844-2_3

[53] Wainaina S, Taherzadeh MJ. Automation and artificial intelligence in filamentous fungi-based bioprocesses: A review. Bioresource Technology. 2023;**369**:128421. DOI: 10.1016/j.biortech.2022.128421

[54] IEA. Renewables 2024. Report [Internet]. 2024. Available from: https://www.iea.org/reports/renewables-2024

[55] Gao J, Yu W, Li Y, Jin M, Yao L, Zhou YJ. Engineering co-utilization of glucose and xylose for chemical overproduction from lignocellulose. Nature Chemical Biology. 2023;**19**(12):1524-1531. DOI: 10.1038/s41589-023-01402-6

[56] Rettenbacher LA, Arauzo-Aguilera K, Buscajoni L, Castillo-Corujo A, Ferrero-Bordera B, Kostopoulou A, et al. Microbial protein cell factories fight back? Trends in Biotechnology. 2022;**40**(5):576-590. DOI: 10.1016/j.tibtech.2021.10.003

[57] De S, Mattanovich D, Ferrer P, Gasser B. Established tools and emerging trends for the production of recombinant proteins and metabolites in *Pichia pastoris*. Essays in Biochemistry. 2021;**65**(2):293-307. DOI: 10.1042/EBC20200138

[58] Grand View Research. Citric acid market size, share & trends analysis report. Food Additives & Nutricosmetics. 2023. p. 65. Available from: https://www.grandviewresearch.com/industry-analysis/citric-acid-market/toc

[59] BCC Publishing. Global Markets for Enzymes in Industrial Applications Report. 2025. Available from: https://www.bccresearch.com/market-research/biotechnology/global-markets-for-enzymes-in-industrialapplications.html

[60] Grand View Research. Biodiesel Market Size, Share & Trends Analysis Report. 2022. p. 65. Available from: https://www.grandviewresearch.com/industry-analysis/biodiesel-market

[61] Good Food Institute. State of the Industry Report: Fermentation. 2024. Available from: https://gfi.org/resource/fermentation-meat-seafood-eggs-dairy-and-ingredients-state-of-the-industry/

[62] FAO. The State of Food and Agriculture. 2024. Available from: https://www.fao.org/publications/fao-flagship-publications/the-state-of-food-and-agriculture/en

[63] Ng YJ, Chan SS, Khoo KS, Munawaroh HSH, Lim HR, Chew KW, et al. Recent advances and discoveries of microbial-based glycolipids: Prospective alternative for remediation activities. Biotechnology Advances. 2023;**68**:108198. Available from: https://www.sciencedirect.com/science/article/pii/S0734975023001052

[64] El-Gendi H, Saleh AK, Badierah R, Redwan EM, El-Maradny YA, El-Fakharany EM. A comprehensive insight into fungal enzymes: Structure, classification, and their role in Mankind's challenges. Journal of Fungi. 2022;**8**(1):23. Available from: https://www.mdpi.com/2309-608X/8/1/23

[65] Sharma K, Kumar P, Singh J. Ecology and diversity of microbial communities involved in the removal of priority contaminants and micropollutants in waste. In: Shah MP, Shah N, editors. Development in Waste Water Treatment Research and Processes. United Kingdom: Elsevier; 2024. pp. 181-199. Available from: https://www.sciencedirect.com/science/article/pii/B9780443136092000148

[66] Liu Q, Gao R, Li J, Lin L, Zhao J, Sun W, et al. Development of a genome-editing CRISPR/Cas9 system in thermophilic fungal *Myceliophthora* species and its application to hyper-cellulase production strain engineering. Biotechnology for Biofuels. 2017;**10**(1):1-14

[67] Zou G, Nielsen JB, Wei Y. Harnessing synthetic biology for mushroom farming. Trends in Biotechnology. 2023;**41**(4):480-483. DOI: 10.1016/j.tibtech.2022.10.001

[68] Grimm LH, Kelly S, Krull R, Hempel DC. Morphology and productivity of filamentous fungi. Applied Microbiology and Biotechnology. 2005;**69**(4):375-384. DOI: 10.1007/s00253-005-0213-5

[69] Garcia-Ochoa F, Gomez E. Bioreactor scale-up and oxygen transfer rate in microbial processes: An overview. Biotechnology Advances. 2009;**27**(2):153-176. Available from: https://www.sciencedirect.com/science/article/pii/S0734975008001079

[70] Marković M-P, Cingesar IK, Keran L, Prlić D, Grčić I, Vrsaljko D. Thermal and mechanical characterization of the new functional composites used for 3D printing of static mixers. Materials. 2022;**15**:6713

[71] Zhang C, Ali Khan RA, Wei H, Wang R, Hou J, Liu T. Rapid and mass production of biopesticide Trichoderma Brev T069 from cassava peels using newly established solid-state fermentation bioreactor system. Journal of Environmental Management. 2022;**313**:114981. Available from: https://www.sciencedirect.com/science/article/pii/S0301479722005540

[72] Wang Z, Zeng D, Zhu Y, Zhou M, Kondo A, Hasunuma T, et al. Fermentation design and process optimization strategy based on machine learning. BioDesign Research. 2025;**7**(1):100002. DOI: 10.1016/j.bidere.2025.100002

[73] Wang S, Chen C, Rzasa JR, Tsao C, Li J, Vanarsdale E, et al. Redox-enabled electronic interrogation and feedback control of hierarchical and networked biological systems. Nature Communications. 2023;**14**:8514

[74] Ha S-J, Galazka JM, Rin Kim S, Choi J-H, Yang X, Seo J-H, et al. Engineered *Saccharomyces cerevisiae* capable of simultaneous cellobiose and xylose fermentation. Proceedings of the National Academy of Sciences. 2011;**108**(2):504-509. Available from: http://www.pnas.org/cgi/doi/10.1073/pnas.1010456108

[75] Zheng X, Guo Y, Chen M, Lu Y, Du Y, Lei Y, et al. Promoter engineering with programmable upstream activating sequences in *Aspergillus niger* cell factory. Microbial Cell Factories. 2025;**24**(1):20. DOI: 10.1186/s12934-025-02642-y

[76] Yang P, Wu W, Chen J, Jiang S, Zheng Z, Deng Y, et al. Thermotolerance improvement of engineered Saccharomyces cerevisiae ERG5 Delta ERG4 Delta ERG3 Delta, molecular mechanism, and its application in corn ethanol production. Biotechnology for Biofuels and Bioproducts. 2023;**16**(1):66. DOI: 10.1186/s13068-023-02312-4

[77] Xia Y, Li Y, Shen W, Yang H, Chen X. CRISPR-Cas technology for bioengineering conventional and non-conventional yeasts: Progress and new challenges. International Journal of Molecular Sciences. 2023;**2023**:15310

[78] Grüttner S, Kempken F. A user-friendly CRISPR/Cas9 system for mutagenesis of Neurospora crassa. Scientific Reports. 2024;**14**(1):20469. DOI: 10.1038/s41598-024-71540-x

[79] Qiao K, Wasylenko TM, Zhou K, Xu P, Stephanopoulos G. Lipid production in *Yarrowia lipolytica* is maximized by engineering cytosolic redox metabolism. Nature Biotechnology. 2017;**35**(2):173-177. DOI: 10.1038/nbt.3763

[80] Qin J, Liu N, Abid U, Coleman SM, Wang Y, Fu Q, et al. Metabolic engineering of *Yarrowia lipolytica* for conversion of waste cooking oil into omega-3 eicosapentaenoic acid. ACS Engineering Au. 2025. DOI: 10.1021/acsengineeringau.4c00053

[81] Xue S-J, Li X-C, Liu J, Zhang X-T, Xin Z-Z, Jiang W-W, et al. Efficient sugar utilization and high tolerance to inhibitors enable *Rhodotorula toruloides* C23 to robustly produce lipid and carotenoid from lignocellulosic feedstock. Bioresource Technology. 2024;**407**:131146. Available from: https://www.sciencedirect.com/science/article/pii/S0960852424008502

[82] Wheeler NE. Responsible AI in biotechnology: Balancing discovery, innovation and biosecurity risks. Frontiers in Bioengineering and Biotechnology. 2025;**13**:1537471

[83] Zhang Y, Chen S, Yang L, Zhang Q. Application progress of CRISPR/Cas genome-editing technology in edible fungi. Frontiers in Microbiology. 2023;**14**:1169884

[84] Almeida DA, Horta MAC, Ferreira Filho JA, Murad NF, de Souza AP. The synergistic actions of hydrolytic genes reveal the mechanism of Trichoderma harzianum for cellulose degradation. Journal of Biotechnology. 2021;**334**:1-10. Available from: https://www.sciencedirect.com/science/article/pii/S0168165621001243

[85] Melnik BS, Glukhova KA, Sokolova EA, Balalaeva IV, Garbuzynskiy SO, Finkelstein AV. Physics of ice nucleation and antinucleation: Action of ice-binding proteins. Biomolecules. 2024;**14**(1):54. DOI: 10.3390/biom14010054

[86] Rhodes J, Jacobs J, Dennis EK, Manjari SR, Banavali NK, Marlow R, et al. What makes *Candida auris* pan-drug resistant? Integrative insights from genomic, transcriptomic, and phenomic analysis of clinical strains resistant to all four major classes of antifungal drugs. Antimicrobial Agents and Chemotherapy. 2024;**68**(10):e00911-e00924. DOI: 10.1128/aac.00911-24

[87] Horta MAC, Steenwyk JL, Mead ME, Braz Dos Santos LH, Zhao S, Gibbons JG, et al. Examination of genome-wide ortholog variation in clinical and environmental isolates of the fungal pathogen aspergillus fumigatus. MBio. 2022;**13**(4):e01519-22. DOI: 10.1128/mbio.01519-22

[88] Naranjo-Ortiz MA, Gabaldón T. Fungal evolution: Major ecological adaptations and evolutionary transitions. Biological Reviews. 2019;**94**(4):1443-1476

[89] Zhao Q, Yang Z, Xiao Z, Zhang Z, Xing J, Liang H, et al. Structure-guided engineering of transcriptional activator XYR1 for inducer-free production of lignocellulolytic enzymes in Trichoderma reesei. Synthetic and Systems Biotechnology. 2023;**8**(4):732-740. Available from: https://www.sciencedirect.com/science/article/pii/S2405805X23000923

[90] Bankevich A, Bzikadze AV, Kolmogorov M, Antipov D, Pevzner PA. Multiplex de Bruijn graphs enable genome assembly from long, high-fidelity reads. Nature Biotechnology. 2022;**40**(7):1075-1081. DOI: 10.1038/s41587-022-01220-6

[91] Gabriel L, Brůna T, Hoff KJ, Ebel M, Lomsadze A, Borodovsky M, et al. BRAKER3: Fully automated genome annotation using RNA-seq and protein evidence with GeneMark-ETP, AUGUSTUS, and TSEBRA. Genome Research. 2024;**34**(5):769-777

[92] Emms DM, Kelly S. OrthoFinder: Phylogenetic orthology inference for comparative genomics. Genome Biology. 2019;**20**(1):238. DOI: 10.1186/s13059-019-1832-y

[93] Danecek P, Bonfield JK, Liddle J, Marshall J, Ohan V, Pollard MO, et al. Twelve years of SAMtools and BCFtools. GigaScience. 2021;**10**(2):giab008. DOI: 10.1093/gigascience/giab008

[94] Blin K, Shaw S, Augustijn HE, Reitz ZL, Biermann F, Alanjary M, et al. antiSMASH 7.0: New and improved predictions for detection, regulation,

chemical structures and visualisation. Nucleic Acids Research. 2023;**51**(W1): W46-W50. DOI: 10.1093/nar/gkad344

[95] Mansourvar M, Funk J, Petersen SD, Tavakoli S, Hoof JB, Corcoles DL, et al. Automatic classification of fungal-fungal interactions using deep leaning models. Computational and Structural Biotechnology Journal. 2024;**23**:4222-4231. Available from: https://www.sciencedirect.com/science/article/pii/S2001037024004008

[96] Picek L, Šulc M, Matas J, Heilmann-Clausen J, Jeppesen TS, Lind E. Automatic fungi recognition: Deep learning meets mycology. Sensors. 2022;**22**(2):633. Available from: https://www.mdpi.com/1424-8220/22/2/633

[97] Hiltemann S, Rasche H, Gladman S, Hotz H, Larivi D. Genome Annotation with Funannotate (Galaxy Training Materials). 2025. Available from: https://training.galaxyproject.org/training-material/topics/genome-annotation/tutorials/funannotate/

[98] Eren AM, Esen ÖC, Quince C, Vineis JH, Morrison HG, Sogin ML, et al. Anvi'o: An advanced analysis and visualization platform for 'omics data. PeerJ. 2015;**3**:e1319. DOI: 10.7717/peerj.1319